KU-736-341

Physiology and the scientific method

E. M. Scott *and* J. M. Waterhouse

D22.

SMS   Biomed

Manchester University Press

UNIVERSITY OF WOLVERHAMPTON
LIBRARY

Acc No.
850294

CLASS 535

571.1

CONTROL

DATE
27. SEP. 1994

SITE
RS

Copyright © E.M. Scott and J.M. Waterhouse   1986

Published by Manchester University Press,
Oxford Road, Manchester M13 9PL, UK
27 South Main Street, Wolfeboro, NH 03894-2069, USA

British Library cataloguing in publication data

Scott, E.M.
    Physiology and the scientific method.
    1. Physiology — Laboratory manuals
    I. Title   II. Waterhouse, James M.
    591.9   QP44

Library of Congress cataloging in publication data applied for

ISBN 0 7190 1898 6 *cased*
ISBN 0 7190 2262 2 *paper*

Typeset
by Williams Graphics, Abergele, North Wales

Printed in Great Britain by
Robert Hartnoll (1985) Ltd
Bodmin Cornwall

Contents

Preface   *page* vii

Acknowledgements   viii

Part I

Chapter 1 Introduction   3

         2 Planning the experiment: choosing the species
                and type of preparation   8

         3 Planning the experiment: experimental design   19

         4 Execution of the experiment: is the preparation
                healthy?   29

         5 Execution of the experiment: the intervention,
                measurement and recording   39

         6 The form, presentation and treatment of results   58

         7 The statistical assessment of results   82

         8 Drawing conclusions   101

Part II

         Introduction   123

Abstract A The effect of plasma albumin concentration
                on albumin synthesis by the liver   125

         B The effect of salt loading on salt glands   137

         C The effect of stimulation of the nucleus ambiguus
                upon the plasma concentration of insulin   150

         D The response of medullary neurones to
                mechanical loads imposed during inspiration   163

Abstract E   The effect of occlusion of the common carotid
arteries (carotid occlusion) on the firing pattern
of hypothalamic supraoptic neurones   178

F   Effects of drink temperature on the emptying
of liquid from the human stomach   188

Index   199

Preface

Any degree course in science must be concerned to a greater or lesser extent with the 'scientific method', that is the means by which hypotheses are formulated and then experimentally tested. However, it is our experience that the student is rarely taught the principles that are involved, but rather left to deduce them for himself from the numerous examples he is given. Moreover, 'experimental design' is often interpreted, not in any wide sense, but rather in the narrow sense of 'statistical design'.

The present book has been written to deal with these problems. In Part I we deal with the different aspects of the scientific method in a systematic way. Thus, we open with an account of why we do experiments and then go on to consider the types of problem associated with the planning and performance of them. We consider the factors that determine, for a particular study, whether *in vitro* or *in vivo* work is appropriate, which species is preferred and the importance of control phases in an experiment. We then discuss the problems associated with the choices of stimulus together with the choice of apparatus required for appropriate measurement and some of the difficulties associated with making recordings. Next we cover different ways in which the results can be presented and calculations can be made from them. We then consider the kind of statistical test that is most appropriate for a quantitative assessment of results and, finally, how to determine if the outcome of the experiments substantiates or refutes the original hypothesis.

In all cases we illustrate the arguments by a wide variety of examples taken from biological sciences in general, but from physiology in particular. These examples are chosen so that the *general principles* are put across and the student can supply detailed examples for himself from the scientific literature he reads. We have assumed that the reader has a knowledge of physiology, at least at the second MB ChB level.

In Part II, the reader is able to test his understanding of the first part by reading six edited 'papers' based very loosely upon published research work. These abstracts cover a range from *in vitro* preparations to whole animal preparations of more direct clinical relevance. With each abstract is a series of questions that is intended to direct the reader's criticism of the abstract. Sample — not model! — answers are provided which stress particular aspects of the scientific method where appropriate.

The book is suitable for undergraduates reading for a degree in biological sciences, particularly in the second and third years. It is also a useful manual for all new postgraduates embarking upon research. In addition, it is a valuable guide to the problems of devising experiments and interpreting results from them and thus should be useful to a wide range of people interested in evaluating the scientific literature including doctors, nurses and dentists.

Acknowledgements

Our sincere thanks to Professor J. L. Cloudsley-Thompson (Zoology, Birkbeck, London) for comments on the final manuscript; Professors M. Case and R. Green, Drs D. Paul, S. Rutishauser, Messrs A. Lewis, S. White (Physiology, Manchester) for comments on earlier drafts; Mr A. Gibbs (Community Medicine, Manchester) for continual advice on all aspects of the statistics; Mrs M. Hagan for her typing; Mrs S. Millward for the diagrams; and to the students who, over the years, have helped us clarify our own minds.

EMS
JMW
Manchester, September 1985

Part I

Chapter 1

Introduction

Reading the newspapers one is often given the impression that scientific breakthroughs happen overnight. This is rarely the case. Most new developments are preceded by many years of work. To take an example, the discovery that the incidence of neural tube defects such as spina bifida in the fetus could be significantly reduced if the mothers were given vitamin supplements before conception and in the early weeks of pregnancy was actually the culmination of several lengthy studies carried out in different centres.

Subsequent chapters in this book will consider the factors which are involved in the planning and execution of experiments and in the handling and interpretation of the data obtained during them. In this chapter we shall consider briefly the ways in which hypotheses or theories are postulated and how scientists set out to test whether or not their hypothesis might be the correct one. The philosophy underlying the formulation and testing of hypotheses is a vast discipline, far outside the scope of this book. However, some indication of the ways in which scientists approach these problems may be useful to someone starting a research project.

1.1 Formulating hypotheses

At the simplest level there are two general approaches to the formulation of hypotheses. In one, data are collected which are relevant to the topic and then a hypothesis is formulated which accounts for that body of data. The second approach is to formulate the hypothesis as the starting point, before carrying out any experiments. The hypothesis then is more of a 'shot in the dark'.

Although in the past there have been great divisions between scientists holding one or other of these two views, these days scientists are greater in their numbers, in the diversity of their views and, in general, in their ignorance of the philosophy of science. Within the scientific community there are some who are both interested in (and knowledgeable of) the philosophy of science, but these are the exceptions rather than the rule.

How, then, do scientists, largely ignorant of the philosophy of science, formulate their hypotheses? The starting point for all must be a desire to find out something, a natural curiosity about a particular subject. To return to our example of the spina bifida trial, the starting point here would be the desire on the part of the clinicians and scientists to find out why this defect arises so that they can help to eliminate such defects and the anguish that they bring to the families concerned. Sometimes the problem seems to be tackled 'because it's there' or 'because it would be interesting to know'.

The next stage is the formulation of the hypothesis. How would the scientists interested in spina bifida have decided that poor maternal nutrition was the cause? They will have observed that the incidence of these disorders varies across the world and also, within this country, between the socioeconomic classes. There are of course a host of possible explanations for the differences in the incidence of these defects between the social classes. It may be entirely genetic, entirely environmental or a mixture of the two. If environmental it might be any of the many possible influences upon the fetus *in utero*; perhaps the mothers who are married to men who work at unskilled jobs eat less (or differently), rest less, smoke more, are more stressed or have poorer living conditions than those married to professional men. How would the experimenter choose one of these factors to see if it is responsible for the fetal abnormality? In this he can be guided by a number of factors. He may, for example, decide on the basis of plausible theoretical grounds that poor maternal nutrition is the cause of the defects. This belief, together perhaps with anecdotal information either from the literature or from his personal experience, will be the basis for his formulating the hypothesis that the high incidence of neural tube defects in some social classes is the result of poor maternal nutrition. In other words, in this case the hypothesis has been devised on the basis of a consideration of several general factors: theoretical grounds, anecdotal evidence, and epidemiological data from first-hand experience or the scientific literature. There are other 'starting points', examples being: the observation of an 'anomalous' result with the desire, to find out why this should have been so; the consideration that principles or findings in one area of research might be transposed and applied to another area; and the role of 'chance'. Chance is often claimed to be the source of a hypothesis in the sense that the hypothesis 'suddenly appeared'. For example, Kekule is said to have dreamed of the structure of the benzene ring, and Watson and Crick were 'shuffling' around models of DNA bases when they suddenly found an arrangement that could form the basis of the structure of DNA. It must be realized that the formulation of a hypothesis 'by chance' nevertheless requires that the hypothesis is *feasible* and scientific expertise is required to decide this. No doubt many other patterns of DNA bases were found by Watson and Crick but these models were rejected or not even considered seriously because they were, biochemically, untenable.

Although sometimes an entirely original hypothesis is formulated by an individual and he can recount the exact time at which the idea came to him, this is unusual. More commonly a group of scientists will be discussing ideas, either alone or with their students, and through such discussions theories evolve. At scientific meetings, groups of scientists meet to compare their results, criticize the techniques and approaches used by others and from such meetings, often later in the day at a social event, ideas are introduced, discussed and perhaps rejected — but they may evolve into a working hypothesis. Once the hypothesis has been formulated it is often difficult in the cold light of the morning to remember precisely whose idea it was!

1.2 Setting about testing hypotheses

Whatever is the means by which a hypothesis originates, what is important is that the hypothesis is capable of being tested. Having formulated the hypothesis, the scientist can then make a prediction which logically follows from the hypothesis and which can be tested. The experimenter must take care to establish firmly the logical relationship between the hypothesis and the specific prediction following from the hypothesis. It is no use spending a great deal of time, effort and money for the purposes of designing and executing a series of experiments to answer a particular question if, at the end of all this, answering that question is not a decisive test of the hypothesis. For example, suppose the hypothesis is that a high-fat diet can result in premature death. Suppose also that it is known that diets rich in fat can cause obesity. To show that obese subjects have a decreased life expectancy in comparison with non-obese controls does not necessarily support the original hypothesis. Obese individuals might differ from controls not only in the size of their fat stores but also in the intake of some other foodstuffs. A decisive test of the hypothesis would require the two groups of individuals to be identical in all factors (including degree of obesity and dietary habits) except for fat intake.

The way in which the hypothesis should be tested is again a matter of controversy. One point of view is that the way to test a hypothesis is to make predictions, verify them and thereby confirm the hypothesis. If one wished to gain support for the hypothesis that a nerve bundle is composed of only large diameter fibres, then each occasion on which a large nerve fibre was found in the nerve bundle would do this. If a sufficient number of observations of large nerve fibres were made then the experimenters might consider the support for the hypothesis to be sufficient to confirm it. But the hypothesis can never achieve the status of 'proved' as might be the case with a geometry theorem or some logical proposition. One can never know that a hypothesis will apply at all times and in all circumstances. However many large fibres were found these results would not refute the possibility that there were also *small* nerve fibres in the trunk. The opposing school of opinion argues that one should set out to falsify or refute the hypothesis. There are advantages in tackling the problem in this way since it is possible to be far more decisive. Thus, if one sets out to refute the hypothesis that all the fibres are large, then the finding of a *single* small fibre would do so.

The process of falsification of a hypothesis is particularly useful when one is looking at two variables and is interested in whether a change in one is causing a change in the other (see Chapter 8). For example, if one had formulated the hypothesis that the increase in ventilation during exercise was caused by a rise in P_{CO_2}, if P_{CO_2} did not rise *before* ventilation increased, then the hypothesis would be disproved. However, if P_{CO_2} did rise before ventilation increased this does not prove that the change in P_{CO_2} is causing the increase in ventilation. In this example, P_{O_2} may have fallen, $[H^+]$ or temperature increased or there may have been an increased input to the respiratory neurones from receptors in the moving joints or from impulses generated in the motor cortex, all or any of which might have produced the increase in ventilation.

Thus, when assessing the status of a hypothesis, often all one can say is that, *up to now*, the hypothesis has not been disproved — so, for the time being, it stands.

To return to our spina bifida example. How could the hypothesis be tested? The experimenters could look at the dietary intakes of various nutrients in the two populations of women, that is, those married to either unskilled or professional men. The observation that the poorer women ate less could be used in support of the hypothesis, but again does not prove it to be correct. (Perhaps the wives of professional men are greedy!) The design of such studies must ensure that the two groups are equal in *all* other respects.

A more decisive test, and one that has been done, is to choose a group of women and give them a nutritional supplement before conception and in the early weeks of pregnancy and to compare the incidence of spina bifida in this group with that in a group of women who did not receive such supplements (but were in all other respects identical). Since neural tube defects are comparatively rare the numbers taking part could be reduced by dealing only with 'at risk' mothers — that is those who have already had one child with a neural tube defect. Studies similar to that outlined above showed that the incidence of neural tube defects in these high-risk mothers could be significantly reduced if the mothers received a nutritional supplement in the early weeks of pregnancy. If such supplements had not significantly decreased the incidence of these defects, the hypothesis would have been disproved.

Even though the results support the present hypothesis, they do not offer an explanation of the result. It might be that the nutritional supplement directly affects some biochemical pathway that must function for normal neural tube development. Or it may be that poor nutrition merely predisposes the fetus to another genetic or environmental factor. Only further hypotheses and tests would help in these matters. However, on a more practical level, if the administration of such nutritional supplements continues to reduce the incidence of neural tube defects, the treatment will become accepted because of its efficacy even if its mechanism of action is unknown. As a result, clinicians will be able to offer advice to those women whose previous children have suffered from spina bifida in the hope that, with appropriate treatment, they can give birth to a normal healthy child. At such times, the 'proof' of the hypothesis and mechanisms of action become academic matters.

In summary, therefore, the scientist needs to devise hypotheses and then to test them by experiment. In subsequent chapters we shall explain how suitable experiments are designed and performed and how the results from them are interpreted.

Chapter 2

Planning the experiment: choosing the species and type of preparation

In this chapter we shall discuss some of the points that an experimenter must consider when planning his experiments. These points apply whether the experiments are designed for the collection of data or to test a prediction of a hypothesis.

2.1 Choice of species

One of the first decisions to be made is which species to use for the experiments. In some studies the very nature of the problem determines the species to be used. For example, one might be concerned in a particular species with the treatment of a disease, the influence of diet or the effects of differing techniques of housing. Alternatively one might want to describe the breeding cycle of a particular insect which carries a disease harmful to man or the effect of a new fertilizer upon crop growth or of a new vaccine against rabies in dogs. Even though many of these studies ultimately benefit man (indeed, that might be the main reason for performing them), the species used in the experiments is not man but is prescribed by the investigations.

In some studies, the animal to be used *must* be man. This applies particularly to the testing of new drugs and the evaluation of new medical treatments. No matter how comprehensive are the studies on animals, the ultimate test must be made on man. Moreover, the clinical studies in man must be carried out on a large number of people over a considerable period of their lives (as in the case of antihypertensive or diuretic drugs, for instance). In spite of the careful regulation of the testing of medicines, cases still occur where drugs that have undergone comprehensive tests using animals turn out to have harmful effects either on the patient or, in some cases, on her unborn child. This is one of the problems when results from one species are extrapolated to another, as will be mentioned again later. For these reasons, in some experiments man is the only species of choice.

2.1.1 *The concept of a biological model*
In many other instances, if the hypothesis to be tested arises from a clinical problem in man or a desire to learn more about the physiology of man, then the preferred animal is man. However, if the experiments involve extensive surgery, such as the removal of an organ, tissue or a part of the brain, or the sectioning of a nervous pathway, then in these cases it is not possible to carry out the experiments using man and another species must be chosen.

Thus in many studies the experimenter has to choose a biological model — an experimental preparation which resembles man (or another species in which he is interested) with respect to the variable being investigated but associated with which are less ethical difficulties. In addition, the model may be technically easier to work with and, on a practical note, it may be considerably cheaper. (Such experimental models must not be confused with the mathematical models which may be used to describe a physiological process; see Section 8.4.) In addition to ethical considerations, there will be several factors that need to be taken into account when the choice of a biological model is made — factors both theoretical and practical.

2.1.2 *Theoretical factors in choosing a model*
If we are ultimately concerned with what happens in man, we must choose an animal which resembles man as closely as possible. This is sometimes taken to mean an animal nearest to man in the evolutionary scale, that is, a mammal if not a primate. However, this is not always the answer. A more important consideration is to choose an animal which resembles man as closely as possible with regard to the system or type of behaviour that is being investigated. Thus, if we were concerned with a response mediated by the vagal nerves, we would choose an animal such as the dog which, like man, has a high level of resting vagal tone; the cat with its higher level of sympathetic tone at rest would be inappropriate. Similarly, in studies on the gastrointestinal tract, the normal diet of the animals needs to be considered when choosing the species.

However closely the species chosen is thought to resemble man, the experimenter must be careful when extrapolating between species.

For example, the reflex inhibition of breathing mediated by stretch receptors in the walls of the airways (the Hering−Breuer reflex) is very pronounced in mammals such as the rabbit but is unimportant in the regulation of breathing in man. Such problems of extrapolation from animals to man are particularly pronounced when one is considering behavioural studies; for example, do studies on the effects of over-crowding on the behaviour of rats really give us much insight into the social and psychological consequences when humans live in high-rise housing?

However, against this 'evolutionary' approach is the view that many biological and physiological processes, such as the transcription of nuclear DNA, the activity of an internal clock-like mechanism, and the physical and thermodynamic laws which govern the flow of fluids through tubes or of molecules down gradients, are *common* to all plants and animals. In such cases, results obtained from simpler animals or plants, even from *in vitro* experiments upon isolated cells or tissues, might provide invaluable material that is of direct relevance to man.

2.1.3 *Practical factors in choosing a model*
In other cases, particularly those that are undertaken to increase our understanding of how a process works, factors other than how much the experimental species resembles man may be more important.

Often there is a species on which many of the previous studies have been carried out and there is a wealth of background data relevant to the proposed investigation. If a *new* species is used then a great deal of preliminary work has to be carried out to establish a background of data in the light of which the new investigation can be carried out. Thus, there is a great deal of pressure on the investigator to select his animal species from one of the small number of animals or plants in common use. For example, most studies on the haemodynamics of the circulation have been carried out in the dog and thus there is a wealth of data obtained from this species; if one were to switch the experimental species, say to the rat, then much of the early work on the dog would have to be repeated using rats.

The tendency to stick with the conventional species is also reinforced by technical considerations. Particular techniques are often developed on a particular species. Our knowledge of the way in which action potentials are generated in an axon has been dominated by results obtained using the giant axon from the squid. In this example, the species was chosen because of its *differences* from man, namely because it contained very large diameter single axons which are very much easier to study than the axons in man which are smaller and arranged in nerve trunks. Also, the early studies on cardiac electrophysiology utilized amphibian hearts, again because technically it was easier to make intracellular recordings than it is in mammalian hearts. A particular strain of rat (the Munich–Wistar rat) is often used for studies in renal physiology since in this strain of rat the glomeruli are found near to the surface of the kidney and hence are more readily accessible to micropuncture techniques.

If an experimenter wishes to change the experimental species from that conventionally used in a particular field then again this would involve a great deal of additional work, in this case, in redeveloping or modifying techniques perfected in another species.

In some studies, it might be appropriate to choose an animal in which a particular response present in man is likely to be exaggerated. For example, if one wanted to study the functioning of the loop of Henle, an animal such as the kangaroo rat, which has a very long loop of Henle and is able to produce a very concentrated urine, might be the animal of choice. Another good example of this is the use of animals such as the duck or the seal to study the diving response. However, problems of extrapolation might exist. In this example, the extrapolation of data obtained from diving animals to man has led to problems since differences have been found in the responses of, for example, a duck during its normal pattern of diving and man during forced immersion of the face in water.

Further examples of animals showing marked responses that have provided useful physiological data are: llamas (information on high altitude); sea-dwelling birds (secretory processes in salt-secreting glands); polar and tropical dwellers (thermoregulation); migrants (orientation and navigation); and aquatic waders (prevention of heat loss by counter-current mechanisms). Often, animals that are unusual in the possession of certain genetic 'defects' are bred specially to study a particular process. For example: genetically obese mice; particular strains of rats with large glomeruli or diabetes insipidus or those prone to develop strokes or hypertension; and Siamese cats (for studies of vision).

Factors such as the ease with which a particular species will breed and can be maintained in good health in laboratories (and increasingly these days the cost of the animals) are also important considerations when choosing a species. For example, many experiments, in which the dog is the species of choice, are carried out using beagles. A very important consideration in the choice of this type of dog is that they live very happily when put in groups in large cages. Commercially, this is very much easier than dealing with a breed of dog which has to be caged separately.

Thus, the existence of basic data and technical expertise gained from a particular species, together with the more practical considerations of cost, ease of breeding, etc. mean that in practice only a small number of all the possible species is used in laboratory experiments.

However, in spite of these considerations, the experimenter must not be overwhelmed by the arguments which encourage him to stick to one of the commonly used species for that particular field of investigation. There may well be experiments for which the species of choice is different. Often the choice of species will be a compromise but at the onset the experimenter should establish which species he would choose in an ideal world. Then, at least, he is aware of the compromises he has made.

2.2 Type of experimental preparation

The next question that arises concerns the type of experimental preparation: should the study be carried out on a conscious animal or on a single cell or cell fragment in a test tube? Again, within this whole range of possibilities, different types of experiments will be appropriate under different circumstances. Experimental preparations are often divided into two types: *in vivo* and *in vitro* preparations. Literally, *in vitro* means 'in glass' and it was originally used to describe experiments carried out in a test-tube. However, the current usage has widened to include in the *in vitro* division all experiments on part of an animal — for example, an isolated perfused organ preparation would nowadays be considered an *in vitro* preparation even if the organ were not removed from the animal. In general, therefore, *in vivo* experiments are those carried out on the whole animal and *in vitro* those on part of the animal. These latter experiments are sometimes termed *ex vivo* experiments.

2.2.1 *Advantages of an* in vivo *preparation*
The advantage of studies on the whole animal is that they allow one to see what actually happens when all the different mechanisms are operating together. Thus, studies on conscious animals, including man, are invaluable if one wants to observe what happens when we stand up, go up a mountain, are subjected to zero gravity during a space

flight or cross time zones. Such experiments are also of importance in pharmacological studies in that they allow the effect of a drug together with the effect of all the compensatory mechanisms to be studied. It is vital to know that, whatever its precise mechanism of action, the drug has the desired *overall* effect when given to man.

Studies on the intact animal allow the experimenter to study higher brain function including behavioural responses which cannot be studied *in vitro*. In addition, even when studying other functions, because of the limited amount of intervention, it is sometimes argued that the preparation is in a more physiological state than an *in vitro* preparation.

2.2.2 *Disadvantages of an* in vivo *preparation*

However, there are disadvantages associated with working on the whole animal: namely, that it is difficult to obtain detailed information about the different stages of a process. For example, in renal physiology, clearance studies in animals can be used to determine the overall handling of a substance by the kidney but give no details as to the sites and mechanisms responsible. The overall clearance of two substances may be equal although the sites and mechanisms of handling within the kidney may be quite different.

A further problem with the studies on conscious animals is that the observations represent the combined effect of the initial intervention and any subsequent compensatory responses. For example, if one studies the effect of giving a drug, are the changes observed a direct effect of the drug or do they result from compensatory responses? When noradrenaline is injected into an intact animal, a bradycardia is observed; by contrast, when noradrenaline is injected into the fluid bathing a piece of cardiac muscle, a tachycardia, is observed. This latter observation is the direct effect of the drug. In the intact animal, as well as producing a tachycardia, the direct effects of noradrenaline include a widespread vasoconstriction mediated by stimulation of α-receptors on blood vessels. This results in an increase in arterial blood pressure which will in turn stimulate the arterial baroreceptors. The reflexly induced bradycardia is of such a magnitude that it outweighs the direct stimulant effects of noradrenaline on the heart rate.

Moreover, the intervention may affect many systems simultaneously and their responses may interact. For example, the infusion of saline into an animal will activate many different groups of cardiac receptors all having different reflex effects and it will also have a number of effects on the kidney. What the experimenter will see is the sum of all these effects and, of course, the final picture may differ between animals.

To investigate a particular hypothesis, it is often necessary to have very accurately defined sites of stimulation or recording. It may also be necessary to interrupt a homeostatic mechanism at a particular stage. This often cannot be carried out in a conscious animal, since there is a possibility that the animal may suffer pain, stress or discomfort. In order to avoid this, the experimenter has to use an anaesthetic or one of a range of preparations in which areas of the brain have been removed, for example, decorticate, decerebrate or spinal animals. In these models, levels of central nervous control have been progressively removed. Such preparations enable extensive preparatory surgery to be performed without the animals feeling any pain. Having done this, it is possible to control the stimulus, perhaps not as well as in an *in vitro* preparation, but better than in a conscious animal. For example, it is possible to stimulate a single group of receptors in the heart by inserting small balloons into the heart and distending them rather than by intravenous infusions which will not only activate many other receptors but which will also effect the kidney. Similarly, it is possible to record the response more directly, for example, by inserting recording electrodes on the sensory nerve and recording the change in the discharge in these nerves rather than the overall reflex effects. Because extensive surgery is possible, it is also possible in an anaesthetized animal preparation to eliminate the effects either of other processes occurring simultaneously or of compensatory mechanisms. In some instances, this could be done by sectioning the nervous pathways involved.

2.2.2.1 *Anaesthetics.* The use of anaesthetics or preparations such as decerebrate animals therefore has advantages, but their use also presents problems. In the spinal animal, the influence of the brain has been removed completely and only spinal reflex mechanisms remain. In contrast, in the decerebrate animal some reflex mechanisms are exaggerated compared to the intact or the spinal animal because of the removal of inhibitory influences. All anaesthetics will affect the central nervous system but their mechanisms and sites of action differ. If an anaesthetized animal preparation is used, then the choice of anaesthetic should also be given careful consideration. Both long-acting and short-acting anaesthetics are available. The long-term anaesthetics have the advantage of being easier to use in that the animal is given an initial injection of anaesthetic whose effects will last for many hours. These types of anaesthetic however have the disadvantage that the depth of anaesthesia cannot be controlled. Short-term anaesthetics are also available. These are invaluable if the experimenter wants to carry out a procedure at the start of an experiment and allow the animal to regain consciousness. These anaesthetics are useful also in experiments where the animal is to remain anaesthetized for many hours. Although these anaesthetics are more difficult to use as the depth of anaesthesia needs to be assessed every few minutes, when the anaesthetics are given in a controlled manner by infusion, a very precise control of the depth of anaesthesia can be attained and a steady-state maintained over a period of many hours.

Different anaesthetics do not affect all areas within the central nervous system equally. For example, α-chloralose tends to preferentially affect the inhibitory pathways rather than the excitatory pathways modulating reflexes. Therefore since animals anaesthetized with α-chloralose have exaggerated reflexes, this anaesthetic is widely used in the study of cardiovascular and respiratory reflexes. In contrast, the barbiturates tend to produce a preparation in which reflex activity is more suppressed and blood pressure is lower than in the conscious animal. The type and level of anaesthesia can also influence reflex effects and, via its effects on the autonomic nervous system, the basal levels of a wide range of physiological variables. One good example where the choice of anaesthetic determines not merely the magnitude of the response but also the direction of the response is the cardiovascular response to stimulation of chemoreceptors in the carotid body. In studies in animals anaesthetized with conventional anaesthetics, such as chloralose or barbiturates, the direct effect of stimulation of chemoreceptors was found to be a bradycardia. However, this direct effect is normally overridden by hypothalamic influences which produce an alerting response including an increase in heart rate and this can be seen in either the conscious animal or in animals anaesthetized with some modern anaesthetics, for example, the steroid anaesthetics. To cite further examples, the choice of anaesthetic can affect the degree of diuresis that can be evoked in an animal, its ability to thermoregulate, or the response of its central chemoreceptors to an acidic cerebrospinal fluid.

As a result of these differences, data obtained from studies in which different anaesthetics have been used may well conflict. In such circumstances, the experimenter cannot be sure which is the 'correct' anaesthetic. The use of the 'conventional' anaesthetic for the particular type of study and preparation he is using will give the same advantages and disadvantages as have already been discussed when the choice of animal was considered. There is the additional point that an investigation of the differences between results produced by different anaesthetics might give valuable information about both the mode of action of the anaesthetics and the process being investigated.

2.2.3 *Advantages of an* in vitro *preparation*

Because of the difficulties associated with work on conscious animals (see Section 2.2.2) and the interpretive problems which result from the use of the anaesthetic agents, experimenters will often study *in vitro* preparations. These preparations have a number of advantages over studies carried out in whole animals if one is interested in studying mechanisms.

Firstly, it is possible to reach the target better, thus the possible problems of absorption of a substance from the gastrointestinal tract, removal from the circulation by the liver or passage across the blood–brain barrier can be surmounted. Moreover, at the cellular level the use of subcellular fragments or organelles removes any consideration of transfer of substances across the cell membrane.

Secondly, using these *in vitro* techniques, it is easier to control the stimulus and measure the response more easily. For example, the use of an *in vitro* preparation of the squid axon allowed electrophysiologists to determine the way in which action potentials are generated in an axon. By taking the nerve out of the animal the chemical environment both inside and outside the cell could be altered. Further, the potential difference across the membrane could be measured and either maintained at a constant value (the voltage-clamp technique) or varied within predetermined limits. Similarly, patch-clamping has allowed us to study the properties of single ionic channels in cells, and studies on isolated glands allow the secretions from the glands to be studied under different circumstances unaffected by changes in the blood flow to the glands.

Thirdly, when cell culture techniques are used, large numbers of identical cells derived from the same cell line can be used. (However, there is evidence that such cells tend to dedifferentiate, possibly related to their no longer being constrained within a whole animal or tissue, and so the extent to which results from them apply to normal circumstances can be debated.)

2.2.4 *Disadvantages of an* in vitro *preparation*

In vitro studies are not without their own problems of interpretation. The major problem is in assessing how normal the preparation remains when it is placed in an abnormal environment. Some aspects of this will be considered in Chapter 4. At the moment we will mention some of the general principles involved.

Removal of an organ or tissue from the body may well alter the properties of the organ. The mechanical properties of hollow organs or blood vessels may alter profoundly when the organs are removed from the body. For example, *in vivo* distension of the carotid sinus is limited by the tissue surrounding the blood vessels, whereas *in vitro*, when this surrounding tissue is removed, the carotid sinus is much more distensible. In this example, the loss of local excitatory sympathetic nervous reflexes may also contribute to the observed increase in distensibility when the organ is removed. Thus, in some experiments, it may be appropriate to maintain an intact nervous innervation of an otherwise isolated organ.

As one proceeds through increasing degrees of isolation of biological material (from a study of isolated organs, through experiments upon tissue slices and tissue culture techniques to studies involving cell organelles or even biologically important molecules, such as DNA) then the natural mechanisms by which biological materials control their environment (Bernard's *'fixité du milieu intérieur'*) are progressively removed. Thus, tissue slices will receive their nutrition and remove waste substances through the incubation medium rather than their normal blood supply. Consequently, they will lose the control of these processes which is normally exerted by autoregulation of their blood supply. Isolated cells are directly exposed to the culture or incubation medium without the cooperative support of other cells to protect them. Moreover, when isolated organelles or molecules of DNA are studied, then the protection afforded by the cell or nuclear membrane has been removed. In these last cases, there is the added disadvantage that we do not know the detailed composition of their normal environment. The general point is that there becomes an increasing responsibility upon the researcher to show that his techniques are not affecting his biological material deleteriously.

It is extremely difficult to predict from *in vitro* studies how the different mechanisms will act together *in vivo*. The demonstration of a particular mechanism in an *in vitro* preparation does not mean that the mechanism normally functions in the intact animal or, even if it does operate, that its effects will not be overridden either by other processes or by homeostatic control mechanisms.

2.2.5 *Merits of complementary use of* in vitro *and* in vivo *methods*
From the preceding discussion, it can be seen that studies using *in vitro* preparations and whole animals each have their advantages and disadvantages. In order to maximize the advantages, the two types of studies should be complementary.

In practice, several research groups are normally interested in a particular problem and there is a tendency for different research workers to carry out experiments, each using a limited range of animals, preparations and techniques in which they have an expertise. Thus, an experimenter will carry out a series of experiments on, for example, an *in vitro* preparation but, by reference to the scientific literature, will be able to relate his findings to those observed in, for example, conscious man. From an examination of results obtained using *in vitro* and *in vivo* preparations, it may be possible to verify that a response predicted from an *in vitro* finding is present in the intact animal. The comparison of different types of study may also yield useful incidental information about, for example, the effects of different anaesthetic agents or of removal of a piece of tissue from an animal.

Chapter 3

Planning the experiment: experimental design

In general, a scientific experiment entails a comparison of the effect upon the experimental preparation of a manipulation with its absence. Thus, in all studies, there will be control groups or phases where the manipulation is absent and experimental groups or phases when the manipulation is carried out.

Regardless of the species and preparation used there are a number of points which the experimenter must consider when planning his study. Care taken in the design of an experiment can avoid many of the difficulties which can arise later when interpreting the data. It is also important to realize that a knowledge of statistical methods can be useful at this stage. The appropriate number of experiments to be carried out will vary depending on the statistical test to be used and whether the measurements are nominal, ordinal or interval (see Chapter 7).

3.1. The random sample

Since it is not possible to test all the members of a population, a sample is used. This experimental sample must be randomly drawn from the population as a whole. This is often difficult to realize fully and consequently the inferences that are drawn from the sample need not (but might!) apply to the whole population. Thus, it is not certain to what extent:

(1) university students represent the population as a whole (with par-
    ticular reference to factors such as age, intelligence, inexperience
    and willingness to volunteer);
(2) segments of ileum represent the whole gut (does all ileum function
    in the same way; are there further differences if samples are taken
    from the jejunum, colon or sphincters?);

(3) frog skin represents amphibian skin as a whole (skin is unlikely to be sampled from extremities or 'unhealthy-looking' areas);

(4) laboratory animals represent that species in general (abnormally healthy/lack of predators/abnormal environment).

3.2 Unpaired and paired experimental designs

In some studies the control sample is independently drawn at random from the same population as was the experimental sample. This is termed an unmatched or unpaired sample. The use of such unpaired control groups assumes that differences between individuals in the control and experimental samples with respect to factors such as age, size, sex, social class or smoking habits (factors which may themselves affect the measurements to be made) will be randomly distributed between the two groups and so will 'average out'. Such a protocol enables statistical tests appropriate for unpaired data to be used but, as will be discussed in Chapter 7, these are less able to detect small changes than tests using data from a matched or paired design. In a paired design, pairs of individual animals (or tissues or cells) are matched for relevant factors such as age, size or sex. The members of each pair are then randomly assigned to either the control or the experimental group.

It is important that the experimenter thinks carefully about which factors he needs to pair. Sometimes, after analysis of the data, it appears that factors other than those which were originally paired may be important. For example, if one were studying the possible effect of a nuclear power station on the incidence of lung cancer in a group of nearby residents, then these individuals would be paired with individuals who do not live near the power station. Age, sex and smoking habits might be obvious factors to pair between the two groups, but it may be found during the study that the smoking habits of the subjects' spouses differ. Since this additional and as yet unpaired factor might affect the results, it might be necessary to perform another study in which the control and experimental groups were paired for this factor also.

In some studies, it is possible to adopt an even better approach and use the same animals or tissues for both the control and experimental group or phases; thus each animal acts as its own control. For example, in a recent study the metabolism of the muscle of a fractured leg that had been put in plaster was measured. In this case, the metabolism of the injured leg was compared with that of the uninjured (and unplastered) leg, which acted as a control. Similarly, two pieces of tissue could be removed, one being the control sample and the other the experimental sample. Consider also experiments which might investigate the effect of a nerve upon the characteristics of the muscle it innervates. It has been shown that, when the nerve to a fast (twitch) muscle is cut and the muscle is reinnervated by a nerve that normally goes to a slow (postural) muscle, then the muscle develops the mechanical and biochemical properties characteristic of a slow muscle. This is interpreted as an effect of the nerve upon the muscle but the possibility of changes due to damage, temporary paralysis, etc. has to be eliminated. This has been achieved, first, by showing that reinnervation by the original nerve leads to a reestablishment of the original properties of the muscle and, second, by showing that reinnervation of a slow muscle with a nerve that normally goes to fast muscle causes the muscle to develop fast characteristics.

Alternatively, the same preparation can be used sequentially for the different manipulations. Thus, the effect of a β-blocking agent upon the nocturnal rise in melatonin secretion by the pineal gland could be investigated by measuring, in the same animal, the rise twice, first in the absence and then in the presence of the drug. This type of experimental protocol is again a paired design. Importantly, the sequence of the tests with and without the drug in different animals should be randomly decided.

Paired designs allow smaller changes to achieve statistical significance than do unpaired designs since they compensate for the variation which occurs between individuals. This concept of not 'missing' small differences between groups (making a type II error) is considered further in Chapter 7.

3.3 Control and experimental phases of an experiment

Ideally, an experiment consists of three phases: an initial, pre-experimental control phase, the experimental manipulation, and the post-experimental control phase.

3.3.1 *The pre-experimental control phase*
The initial control phase takes place at the start of the experiment. It enables the preparation to 'settle down' allowing, for example, the initial effects of anaesthesia or surgery to wear off or for the subjects to familiarize themselves with the apparatus to be used in the experiments.

Sometimes this phase can last far longer and is a 'run-in' period which is required to standardize all preparations with respect to dietary and fluid status or to allow washout of drugs. This control phase, which is discussed further in Chapter 4, also enables the experimenter to make measurements to establish that the preparation is 'healthy' and that the control values are within normal limits.

The pre-experimental control phase often involves studying the preparation 'at rest', but this need not always be so. Thus, different parts of the experiment might have their own control phase, not all of which are literally 'resting' but rather are the baseline conditions against which the experimental change can be measured (see Chapter 6). In addition, the conditions for the initial control phase can be very strictly defined and standardized—one is 'calibrating the preparation', in effect. For example, if one is performing metabolic studies, then standard conditions with respect to fasting, posture, environmental temperature, humidity and airflow have to be adhered to. Also, in investigations of sensation (particularly visual and auditory response curves), it is of the utmost importance to standardize the ambient conditions.

3.3.2 *The post-experimental control phase*
Ideally, control readings should also be taken after the experimental intervention. However, there are some circumstances when this is not possible as would be the case after ablation of an area of the body or after nerve section, for instance. This second control phase will enable the detection of any changes in baseline levels, for example due to ageing, deterioration of the preparation, changes in the level of anaesthesia or the cumulative or interactive effects of drugs. If control values alter during the course of an experiment it will require careful thought to decide the most appropriate means to assess changes produced during the experimental phase (see Chapter 6). In addition, such changing baselines may imply deterioration of some sort and the experimenter must attempt to minimize them. Clearly, it is advantageous if the experimental procedure can be reversible, an argument for the use of reversible local anaesthesia, or rapidly metabolized drugs where appropriate.

3.3.3 *The experimental phase*
Between the control phases is the experimental phase, in which the manipulation designed to test the hypothesis is performed. Details of some of the problems arising in this phase of experiment are given in Chapter 5. Sometimes, there is more than one experimental manipulation

and this necessitates more than one experimental phase. For example, one might want to stimulate a nerve at different frequencies, to infuse saline at different rates or to perform exercise at different work-loads. In these cases, there should always be a control phase dividing the experimental phases. It is also important that the experimental phases be carried out in random order to avoid cumulative effects of any manipulation or drug or the effects of fatigue, etc. However, such protocols might become unacceptably long with the likelihood of a severe deterioration in the preparation being used. The most suitable solution is to design the experiment so that matched groups are used. Each group would then have only one experimental manipulation carried out. One-factor analysis of variance could then be employed as the statistical technique to analyse the data (see Chapter 7). In the examples given, the 'one factor' that is being considered is the frequency of stimulation, rate of saline infusion or work-load during exercise.

The order in which the experimental phases are conducted is particularly important when comparing the effects of drugs and leads to the crossover type of trial in which drug A is given first with appropriate pre-experimental and post-experimental control phases followed by drug B and its control phases. The sequence is then reversed (drug B before drug A) and the effects compared. This protocol enables a correction to be made if the activity of the second drug is modified by a residual effect due to prior treatment with the first drug.

3.4 Control experiments

Regardless of the type of intervention used the experimenter may often be open to the criticism that he had unwittingly produced another difference or series of differences between his animals during the control and experimental phases in addition to the intervention that he so carefully designed. For example: when he injected the drug he caused, in addition, a change in the pH of the blood; or, in an experiment which investigates the role of a nerve, it was not sectioning the nerve which affected the response but rather the initial surgery which was necessary in order to reach the nerve.

These lines of criticism can often be applied to poorly designed experiments and they are countered by performing a suitable control experiment. The aim of the control experiment is to treat another group of animals or preparations in the same way as the experimental group in all respects except that the experimental manipulation is not performed.

The effect of the experimental manipulation can then be judged by comparing control and experimental groups at the manipulation stage of the experiment. Clearly, to be of value, the control group must be as close as possible to the experimental group during the control phases that exist both before and after the experimental phase. If the groups differ before the experimental phase, then the control group is useless as a control; if the groups differ after the experimental phase this implies some longer-lasting effect of the experimental manipulation.

3.4.1 *Types of control experiment*

The control group is of equal importance to the experimental group when the results are analysed. Its format depends very much upon the type of experiment being performed. In studies lasting many years (see Section 3.6), a necessary control is of a group in whom the changes due to ageing can be measured. If one is investigating the effects of a supplement to the diet, both control and experimental groups would receive the same diet throughout the course of the experiment apart from the supplement which is given to the experimental group. In this case, the control group receives no experimental manipulation but, in other cases, the control group may have a manipulation performed upon it during the experimental phase. If one is interested in the effects of an injection of a drug, then the control group would receive injections of the solution in which the drug was dissolved, 'the carrier', checking that other relevant factors such as the pH, volumes and rates of infusion of the two solutions were the same. In this way, the effect of the drug could be distinguished from that of the carrier (ideally, of course, the carrier should have no effect). In the case of an experiment designed to measure the effects of removal of part of the gut it is necessary to show that any effects were not the result of anaesthesia or general surgical trauma (the experiment itself would not be performed until the animal had recovered from the operation). Here the control group would be anaesthetized and undergo a similar amount of surgery. Strictly, the section of gut should be removed *and then replaced* but this might cause complications in interpreting the results if the piece of gut did not recover normal function. Such a control group is called a 'sham operation' group. In some cases, the preparatory surgery renders the animal abnormal in so far as it differs from unoperated controls due to the removal of some tonic influence. Thus, if one were investigating the importance of the thyroid gland in enabling an animal to combat cold stress, in a paired experimental design it would be necessary to compare animals' responses to cold before and after thyroidectomy. Alternatively, one could compare two groups of animals, one with (control group) and the other without (experimental group) intact thyroid glands. The problem that arises is that thyroidectomy changes the basal metabolic rate so that the animals are no longer identical with respect to initial thermal status. When such changes in baselines values are produced there are difficulties of interpretation when quantitative comparisons between control and experimental groups are being made (see Chapter 6).

The choice of an inappropriate control has been the downfall of many experiments.

3.5 Subjective assessments

In many experiments upon human volunteers, the subjects are required to answer questionnaires or assess their feelings or the severity of their symptoms. Such experiments raise further problems because the subjects often wish to 'aid' the experimenter by telling him the answer they think he wants to hear! So, if one is testing a new drug, the subject (who anyway wants the drug to improve his condition) may — even unwittingly — give the researcher an over-optimistic account of the drug's efficacy.

To avoid this kind of problem, such studies should be carried out so that the subjects do not know if they are receiving the new treatment or drug to be tested, an alternative, well-tested drug or treatment, or a tablet containing, for example, cellulose — the 'placebo'. Even though it is intended that the placebo represents 'no treatment', it may itself have an effect; for example, if cellulose is given, this may be beneficial as it adds to the fibrous content of the gut. Cases have been recorded where, after a clinical trial, a subject has asked the manufacturer to continue supplying him with the placebo as it improved his condition far more than any of the available treatments! Such requests are made after the patient has been told that he has been receiving 'no treatment'. In the case of drugs it is preferable (but not always possible) that all medications look and taste identical. When either the subject or the assessor does not know which treatment has been received, this is termed a *single-blind trial*. Unfortunately, in spite of these precautions, the subjects may deduce that they are in the experimental group if you change their diet drastically or their working habits, for example. In some trials, neither the subject nor the assessor knows whether a particular subject has received a particular treatment until after the experiment. This is termed a *double-blind trial*. It has the advantage that the assessor also cannot bias the outcome of the study by the way in which he asks questions or assesses symptoms.

There are, however, ethical problems associated with such medical trials since it may be necessary for a doctor to know what treatments a patient is getting so that he can look out for harmful side-effects. This criticism applies particularly to double-blind trials, of course. The initial testing in man of a new drug would never be double-blind for this reason. Even for single-blind trials, the experiment may have to be abandoned prematurely if it is felt that by giving the placebo, the patient is suffering as a result of being denied the best treatment available.

3.6 Transverse and longitudinal studies

In some experiments, one is interested in the way in which the effects of a factor changes with duration of exposure, the long-term effects of radiation, drug therapy or shiftwork, for instance. One means of achieving this is to take a cross-section of the population. Within this sample, there will be individuals who will have been exposed to the factor (radiation, therapy, shift work) for different lengths of time. One can then investigate the relationship between length of exposure and any variables such as existence or severity of certain symptoms, as chosen by the experimenter. Such a transverse study suffers from the disadvantage that it must assume that the members of the group are alike in all respects — including their memory and truthfulness — except for the duration of exposure to the factor under consideration. This need not be so, since those who react most adversely to the factor might be more prone to die, to stop therapy or to leave shift work. As a result, the individuals in the sample who have been exposed to the factor for a long time might be unrepresentatively hardy or tolerant. There may also be differences between the groups who volunteer for such studies and those who refuse, a problem that has been raised before (see Section 3.1).

A possible solution to this problem is a retrospective longitudinal study by which the *same* individuals can be studied repeatedly. (There is an analogy here between transverse and longitudinal studies and unpaired and paired protocols as described above.) One such study which might be applied to humans would require individuals to remember their symptoms as they existed on a number of previous occasions. In this way, changes in individuals with time of exposure could be investigated. However, even more than with the transverse study described above, this approach relies upon the correctness of the memory of the individual and upon the assumption that details of the factor being investigated have not changed. Thus, if the amount of protective clothing worn, the drug regime or dose, the number of hours worked or the shift system had changed, then interpretive problems would arise. Further, the severity of symptoms necessary for hospitalization or medical treatment might also have changed over the years. A means to overcome this problem is the prospective longitudinal study. In this, the same individuals are studied on successive occasions. The variables and their means of measurement can be standardized and an exact record of exposure to radiation, drugs, shift work, etc. can be kept. This does not rely upon the memory of the individuals (but still upon their truthfulness!). Clearly, such a study is time-consuming and costly and can suffer as a result of individuals failing to attend subsequent sessions. This loss of data as a result of volunteers failing to complete the study can be a serious problem when the experimenter comes to analyse the data; particularly with analysis of variance (see Section 7.1.1). It cannot be assumed that this group represents a random sample of the whole group — it may well be that they are those most, or least, affected by the intervention. Again, changes in the factors being investigated can cause difficulties but at least an accurate record of them will be made. Further, if the symptoms being investigated are expected to occur in only a small proportion of people, then the total number which has to be studied in prospective studies becomes enormous.

It will be noted that all these repetitive studies have in common the requirement for a control experiment that enables the effect of age to be taken into account.

3.7 Conclusions

We have attempted to explain in these two chapters the best way in which to design the experiment, but might have been guilty of advocating a 'counsel of perfection'. In practice, there are occasions when these recommendations cannot all be fully implemented. This need not render the experiment worthless — indeed, if that were the case, few experiments would be performed! — but it can become an important factor when the statistical analysis is performed and the results are interpreted. Thus, differences between the results from research laboratories can often centre on the assumptions and choices that have been made in these planning stages.

Chapter 4

Execution of the experiment: is the preparation healthy?

4.1 Prior to the experiment

Regardless of the preparation chosen it is absolutely essential, both before the experiment starts and during the experiment, that the preparation be maintained in a state which is as near to physiological normality as possible. If this is not done, then any experimental intervention imposed by the experimenter will be upon a physiological background that is very different from that of the normal resting animal. Not surprisingly, an intervention under these conditions may yield results which differ from those obtained when the animal is in the true resting state. For example, differences in the relative amounts of vagal and sympathetic tone present in experimental animals used in different studies is a very likely explanation for the many different results obtained when the effects of stimulation of baroreceptors on cardiac output are examined.

It may sometimes be difficult to define precisely the limits of physiological normality but, where possible, reference can be made to the values obtained in conscious but resting specimens. Some of these data are available in the literature but on some occasions, especially if work upon a species that is infrequently used is planned, it may be necessary for the experimenter to obtain them for himself.

4.1.1 *Studies on man*
Let us start with the problems associated with working on conscious humans. The main problem here is to ensure that the subject is both in a resting condition and is unstressed. This takes a great deal of time and practice on the part of the experimenter. It is essential that the subjects are reassured at the start of the experiment and that the experimenter explains to the subject *before* starting the experiment the procedures to be carried out. The more the subject can be familiarized with the

procedure, the less stressful it will become. Even a very simple non-invasive technique such as the measurement of blood pressure by using a sphygmomanometer may be very frightening for an uninformed subject and it is known that measurements made before the subject is thoroughly familiar with the apparatus and experiment give erroneous results. For this reason, it is common to ignore the early results from a study.

Thoroughly familiarizing the subjects with the procedures to be adopted in the experiment may also minimize or eliminate any changes that arise because of an effect of training. Consider the example in which the subject is required to perform a task (such as a series of mental tests or a test requiring manipulative skill) on a number of occasions. Initially, he will be improving his performance very considerably as he gains familiarity with the problem and these practice sessions should be finished before the experiment starts. (Of course, by contrast, if it is the rate of acquiring the skill which is being investigated, then the preparative work consists only of explaining fully what is required of the subject.)

It is also important to allow the subjects to rest for a period before starting the experiment. They may well have rushed to be at the laboratory by the appointed time and thus may be recovering from a period of exercise or be generally 'unsettled'. Often control readings are continued until the subject is seen to be in a steady state at the start of the experiment. These factors must be taken into account in the design of the experiment (the pre-experimental control phase, see Chapter 3). Depending on the study, a whole host of other factors may be important, for example, when and what the subject last ate or what drugs he is taking. Thus, in some studies, there may be the necessity for a control period of several hours or even days before the study can begin in which, for example, the intake of food and water of the subject is monitored.

4.1.2 *Study on animals* (in vivo)

If the experimenter intends to use conscious animals then he must prepare himself for a period of weeks or perhaps months in which to familiarize his experimental animals with the procedures to be undertaken. During this period, the animals are normally brought up to the laboratory and gradually introduced to the procedures that will be undertaken in the final experiment. In some species, for example rats, it may be necessary to start this process of familiarizing the animals immediately after birth by ensuring that the animals are handled whilst in their breeding cages.

As in studies on man, it may also be important to control various aspects of the animals' environment. Studies have shown that a wide variety of environmental factors such as the temperature, humidity, air movement, light and sound, can produce both physiological and behavioural changes in animals. The changes produced are diverse including changes in metabolic rate, concentration of hormones, patterns of sleep and wakefulness, growth and lactation.

There are examples in the literature where environmental factors have affected the outcome of experiments. For example, in studies on the toxicity of isoprenaline in rats it was found that if the animals has been exposed to cold before the test then the drug was more than 1000 times more toxic than it was in a control group which had not been exposed to cold. The light/dark and feeding schedules in operation before the experiment begins will have a decisive effect in those cases where circadian rhythms are important, for example in the release of hormones such as cortisol.

These considerations about the care of the experimental animals before the experiment also apply to animals that are going to be anaesthetized for the whole of an experiment. Again, rats seem particularly sensitive to the conditions under which they are kept. For example, very minor changes in diet, or the presence of noise caused by building work outside the animal house, moving the animals' cages or even the introduction of unfamiliar personnel in the animal house can often produce quite significant effects on the experimental results obtained. Such changes may sometimes explain why an experimental result found consistently in a group of animals can no longer be produced in other animals — many experimenters will give accounts of wasted weeks or months while they search for a reason for the sudden failure of what promised to be a successful series of experiments.

It is important to avoid stressing the animals either before or during induction of anaesthesia. There are sound physiological reasons for this in addition to the experimenter's desire to avoid causing the animals any distress. When stressed, the concentration of a number of hormones, for example, adrenaline, renin and cortisol, will increase drastically. In studies on rats, disturbances of the endocrine system were detectable within 1 min of touching the animals' cages and persisted for at least 35 min. There is also likely to be a widespread increase in the activity in the sympathetic nerves resulting in effects on many different organs and tissues; for example, the rate and force of contraction of the heart will increase, there will be a constriction of the arterioles in many regions which will increase blood pressure and the motility of the gut will be reduced. Ventilation, too, is likely to be increased. As with the endocrine changes, these effects are likely to persist for some time.

4.1.3 In vitro *studies*

The above arguments are also important in *in vitro* studies. Biological material that is removed from an animal which was stressed is unlikely to be the same as that removed from an unstressed animal. This is because stress prior to removal of the sample will affect the discharge of nerves and release of hormones. An additional problem is that the surgery required for removing the sample will sometimes be major and traumatic even in expert hands. Moreover, the isolated sample will be less able than a whole animal to correct adverse effects either by homeostatic mechanisms or by metabolic changes as discussed in Section 4.2.2.

4.2 **During the experiment**

Care must be taken to maintain both the control and experimental preparations in a state which is physiologically normal throughout all phases of an experiment. What, then, are the measurements that we can make during the control phases of an experiment to assess the physiological normality of the preparation, why might they change, and how can these variables be controlled?

4.2.1 In vivo *studies*
4.2.1.1 *Level of anaesthesia.* The importance of choosing a type of anaesthetic that is appropriate for a particular series of experiments has already been considered (see Section 2.2.1). It is equally important that attention be paid to maintaining the animal in a steady state of anaesthesia. Changes in the level of anaesthesia may suppress some responses and facilitate others. Moreover, changing the animal's condition during the control phase might result in differences in both the magnitude and even the direction of a response during the experimental phase.

To date, there is no quantitative technique available that can be used to assess the level of anaesthesia induced by all the different types of anaesthetics. However, over the years many experimenters have gained experience in assessing the depth of anaesthesia by monitoring variables such as heart rate and blood pressure and by looking at the magnitude of reflex responses such as the blink reflex, the reflex withdrawal to a noxious stimulus such as pinching a paw, or the reflex contraction of muscles in the trunk in response to a sharp tap on the table. More sophisticated techniques such as assessing the magnitude of evoked responses in the brain have also been used.

4.2.1.2 *Temperature.* Many anaesthetized animals are less able to regulate their body temperature than conscious animals. Moreover, trauma during minor surgery will suppress the hypothalamic thermo-regulatory control mechanisms. Experimental animals may also be subjected to additional heat or cold stresses. For example, if a large incision is made in the chest or abdomen, the heat loss will increase or conversely the intense lighting often used during surgical procedures may result in an increase in heat load. Small animals, because their surface area-to-volume ratio is high, are particularly prone either to lose or gain heat depending upon the environmental temperature.

It is important to maintain the temperature of the experimental animals within normal limits, since changes in body temperature will modify many physiological responses. For example, in addition to the general effects on body metabolism, changes in the inputs from temperature receptors will modify cardiovascular reflexes, the effects of drugs are often temperature-dependent and changes in body temperature may also alter the levels of circulating hormones, thus again altering the basal state of the animal during the control phase of the experiment.

Body temperature can be easily monitored in any experimental animal by using a thermistor introduced into the oesophagus or rectum. This should be routinely done in all experiments. The animals' body temperature can be controlled by the use of heating lamps above and below the operating table.

4.2.1.3 *Ventilation and acid—base status.* It has been observed that during anaesthesia both man and animals become acidotic. The rate and patterns of ventilation may also alter during the course of an experiment as the level of anaesthesia alters.

It is of importance to maintain both the rate and pattern of ventilation and the pH of the blood within normal limits. Changes in the partial pressures of carbon dioxide and oxygen in the blood can have important effects on the body. In addition to producing reflex effects via stimulation of the chemoreceptors, hypercarbia and hypoxia will also modify the discharge of the sympathetic nerves and affect the sensitivity of reflexes, for example, that of the baroreceptor reflex.

Changes in the rate or pattern of breathing, by altering the discharge of receptors in the airways of the lung, will not only modify the ventilatory pattern further, but will also modulate the responses to stimulation of other groups of receptors, for example baroreceptors. Changes in the pH of the blood also have important effects. Acidaemia suppresses the reflex changes in heart rate which result from stimulation of stretch receptors in the atrium and which are mediated by the sympathetic nerves. These changes result partly from changes in the discharge of the nerves, but are also at least partly mediated by changes at the effector since acidaemia also suppresses the effects of electrical stimulation of the sympathetic nerves. Conversely, acidaemia potentiates the effects of electrical stimulation of the vagal nerves.

In the anaesthetized animal, ventilation can be controlled by artificially ventilating the animal. Periodically, the partial pressures of oxygen and carbon dioxide can be measured and the ventilation altered so that these values are maintained within normal limits. A respiratory acidosis can be corrected by increasing the ventilation and a metabolic (or non-respiratory) acidosis can be corrected by administering sodium bicarbonate to the animal. Because of the presence of stores of bicarbonate in the body, it is not possible to predict precisely how much bicarbonate is needed to correct a metabolic acidosis, although attempts to do this are made.

In practice, the acidosis is best corrected by administering some bicarbonate and then measuring the pH of the blood again to see if any further correction is necessary. However, if small animals are used in the experiments, then this procedure may not be possible as the removal of blood samples will represent a significant haemorrhage.

4.2.1.4 *Blood pressure and heart rate.* In many experiments, blood pressure and heart rate are routinely measured. In some studies, blood pressure is measured and used as an index of the adequacy of perfusion of an organ, the experiment being stopped if blood pressure falls below a predetermined level. This can be a useful approach, but the disadvantage with it is that a normal blood pressure may not necessarily be associated with a normal blood flow to an organ if the vessels supplying that organ are markedly vasoconstricted.

The simultaneous measurement of blood pressure and heart rate can yield a great deal of information to the experienced experimenter, including changes in the depth of anaesthesia, the blood volume and the adequacy of ventilation.

4.2.1.5 *Other variables.* Depending upon the purpose of the experiment, other variables may need to be monitored. In experiments where extensive surgery is required, a particular check should be kept on estimates of fluid balance (as assessed from measurements of blood volume, haematocrit or plasma osmotic pressure) and of the plasma concentrations of hormones, particularly catecholamines and adrenal corticosteroids, which are released during trauma.

4.2.1.6 *Viability of the preparation.* In addition, particular care should be given to establishing that the organs or systems being experimented upon are viable. For example: on studies of the liver, that the rate of production of bile and its composition are normal; on salivary glands or kidneys, that the secretion of saliva or excretion of urine is of normal volume and content; and, when studying parts of the brain, that its electrical activity is normal.

It is also important to demonstrate that the preparation is capable of responding in the normal way to stimuli or stresses aimed at the particular organ or system under consideration. Thus an experiment testing some aspect of hypothalamic function might require it to be shown that endocrine responses to neural and humoral stimuli are normal or, if the experiment is upon gut musculature, that gut motility and reflexes are working satisfactorily.

4.2.1.7 *Site of measurement.* In other studies, it may be the concentration of a substance or the partial pressure of a gas *at a particular site* which is important. Such measurements pose more problems than merely assessing the state of the blood in general, but recent technical advances have made some of these assessments possible. For example, the concentration of a number of substances in the cerebrospinal fluid, or the partial pressures of gases at different sites through the wall of a blood vessel, can be measured. Electrodes are also available with which the concentration of ions, including hydrogen ions, within cells can be measured.

Often such studies are carried out as a pilot study at the start of a series of experiments rather than preceding each individual one. This is because they are too time-consuming to permit the planned experiment to follow. They are a necessary proof that the preparation is suitable for the type of experiment that will be performed and are, in fact, to be considered as another type of control experiment.

4.2.2 In vitro *studies*

Factors such as the osmotic pressure, ionic composition, temperature, partial pressures of oxygen and carbon dioxide or pH are important in *in vitro* studies also — and perhaps even more so than in *in vivo* studies as homeostatic mechanisms are decreased or absent. As was discussed earlier, changes in temperature, for example, will alter the rate of metabolism of a tissue. Thus, *in vitro* studies will often be carried out at low temperatures precisely for this reason or because organs and tissues are less susceptible to ischaemia under these conditions. However, problems may arise, for example, in investigating the effects of some drugs since these effects may vary with temperature. Regardless of the temperature selected, it must be maintained at a constant level throughout the experiment.

An adequate supply of oxygen to the tissue is also important, and again it is the amount of oxygen reaching the cells that is important rather than merely the oxygen partial pressures or concentrations in the fluid surrounding or perfusing the tissue. There is, therefore, a limit to the thickness of a tissue which can be used in *in vitro* studies if it is to receive an adequate supply of oxygen. Again, the pH of the surrounding fluid must be measured and controlled. The use of a fluid having a pH that is outside the normal range will have both direct effects on the tissue and also may have indirect effects by, for example, altering the concentration of ionized calcium.

With isolated perfused organs, particular consideration should be given to the composition of the perfusate, its rate of perfusion and the warning signs that deterioration of the organ is becoming marked. Perfusion rates should be similar to those found *in vivo* and in some studies it may be necessary to mimic the normal pulse wave occurring in the vessels being perfused. Inappropriate perfusion rates might be indicated by the presence of abnormal concentrations of a substance in the effluent. Successful organ preparations 'recover' from the immediate trauma of surgery and cannulation, etc. but, ultimately, deterioration becomes progressive. It can be recognized by a number of factors including changes in electrical activity (in suitable organs), alterations in the pressure of the perfusion system or changes in the concentration of, for example, haemoglobin, calcium and glucose in the perfusate.

A detailed discussion of the assessment of the health of cell cultures and suspensions of membrane vesicles, cell organelles and biological molecules is not appropriate here, as it would require detailed biochemical knowledge. However, it is important to realize that the separation techniques that are generally used — the use of hyaluronidase and strong urea solutions to separate cells; homogenization procedures to produce membrane vesicles; free-flow electrophoresis to separate luminal and contraluminal membranes from epithelial cells; the subjection of organelles and large molecules to strong acceleration forces in 'alien' environments of solutions of concentrated caesium chloride, polysaccharides or iodinated derivatives of benzoic acid during centrifugation — might all alter normal biological activity. In such cases, the best advice is to attempt to duplicate results with a variety of approaches on the grounds that it is unlikely that different techniques would produce the same artifacts of preparation.

In tissue culture studies, not only must the experimenter be concerned with controlling the environment so that the cells remain healthy, but also in maintaining an environment in which the cells will differentiate in the normal way.

4.3 Conclusion

In conclusion, the experimenter should give considerable attention to ensuring that the preparation he is using is as near to a physiological one as possible. Again, as in previous chapters, we might be guilty of setting unrealistically high standards for the experimenter. Fortunately, even if such standards are not fully achieved, such is the resilience of living tissue that useful results can still be achieved in spite of everything. It is the case that some physiological functions are hardier than others so that, in many cases, the variable under investigation might be substantially normal even if other, more susceptible, variables show signs of deterioration.

But how can he answer criticisms from other workers that his preparation was either abnormal or not viable? Criticism concerning the abnormality of the preparation can be answered by measuring a number of variables as considered earlier and showing that these values lie within normal limits. The viability of the preparation can be demonstrated by showing that the preparation in question responds in the predictable way to other stimuli, be they drugs, stimulation of receptors or electrical stimulation. Questions concerning the viability of a particular preparation often arise when an experimenter fails to demonstrate a response. The problem then arises as to whether this lack of effect is real or is merely because the preparation was incapable of responding. Evidence to support the former conclusion would be provided if, for example, in the event of failing to show any effect of a substance on a tissue, that tissue was seen to respond to other substances. Similarly, in the case of failure to demonstrate a reflex, other reflexes mediated along the same pathways could be shown still to be intact. However, such tests of the viability of a preparation are not foolproof (see above); some functions seem to be more easily lost than others. Thus, even if an experimenter is himself convinced of both the normality and viability of his preparation, it is worth his while to provide independent evidence to convince the sceptics.

Chapter 5

Execution of the experiment: the intervention, measurement and recording

Experiments consist of an intervention that is designed either to stimulate or remove a system, and a recording that enables the effects of the intervention to be observed.

5.1 The intervention

5.1.1 *General aspects of any intervention*
As will be considered in the following sections, there are many different types of intervention. The intervention may be: stimulation of the whole body (or a part of it); blockade of, for example, a nerve trunk; or ablation of a part of the body, possibly with some type of replacement. But, regardless of the precise type of intervention to be employed, there are a number of general principles which should be considered. Thus, the intervention must be suitable for the particular preparation which is to be used. (In fact, this is the 'choice of species' problem seen from the opposite viewpoint.) Secondly, the intervention must be shown to achieve the task it was required to perform. That is, the intervention stimulated a particular organ or reflex under consideration, sectioned the intended nerve trunk or removed the appropriate area of the brain. As will be discussed later, in some studies it may be necessary to localize the stimulus or site of blockade very accurately so that the intervention can be specific enough.

Just as the resting conditions of many preparations must be standardized, so too it is important that the intervention be standardized between the different experiments. This may be simply done, for example, by giving a consistent dose of a drug (or glucose, in the glucose tolerance test), but if, for example, removal of the drug by the liver varies between individuals, then it may be necessary to standardize the interindividual plasma concentration of the drug. Some of these problems can be overcome by giving a standard dose in terms of 'per kg body weight'. However, matters are not always that simple, as will be discussed in Section 6.1. In the example cited above, it may be that in two individuals of the same body weight their ability to metabolize the drug varies greatly. As a further example of standardizing a stimulus, in some neurophysiological studies it may be important to show that the experimenter is stimulating the same group of neurones in each case. In this case, the stimulus and its spread may be standardized by measuring the amount of current spread rather than merely by controlling the voltage of the stimulus.

Occasionally, such 'standards' have been found to be inadequate and required refinement. For example, original attempts to assess a circadian rhythm in mental performance used repetitive testing with a series of 'standard' tasks measured under 'standard' conditions. However, it was later found that different tasks gave different types of rhythm over the course of the day. We now know that the rhythms of tasks with a high memory-load (remembering strings of random digits, for instance) are inversely related to the rhythm of body temperature, that the rhythms of tasks with a high 'throughput' of information (searching for 'wrong' signals in a stream of data) are directly related and that the rhythms of tasks with both components (mental arithmetic) show an intermediate relationship. The important point is that, once again, anomalous results can lead to a better understanding of the biological processes involved and, in this case, a better 'standardization' of stimuli.

Another consideration (which again will be discussed in detail in the following sections) relates to how faithfully the intervention resembles the normal physiology of the animal.

5.1.2 *Stimulation*

The experimenter may be interested in describing the physiological response to a stimulus, such as a change in posture, day length or diet. Alternatively, the stimulus may represent that found in a less normal circumstance, for example, a period at high altitude, zero gravity, or of illness or trauma. The stimulus may be applied externally to the whole animal, as in the examples given above or, in other types of investigation, the experimenter may wish to examine the physiological effects of a stimulus to a more limited region of the whole animal, for example, to a group of receptors or neurones. In the case of an *in vitro* preparation, the stimulus may be applied to an area as limited as a tissue, cell or cell fragment. Thus one might want to inject a substance into a particular part of the brain, perfuse an organ and change the composition of the perfusate, or look at the effects of calcium on different cellular organelles. Such approaches show clearly the increase in localization of stimulation that is possible, but always at the possible risk of rendering the stimulus 'non-physiological' (see below, p. 42).

If the experimenter is interested in observing the effects of a stimulus which affects the whole body, then the choice of intervention and the precise way in which it is executed will usually present few problems. The problems with this type of experiment arise later, firstly in trying to predict from the final observations the mechanisms producing the changes and then in trying to design further experiments to investigate in more detail the role that these different mechanisms are playing in producing the final response. Problems more usually arise when the investigator is trying to produce a more limited stimulus. Thus he must then provide evidence that the technique he is using is capable of stimulating, for example, the particular group of receptors or the neurones in the brain or the secretory tissue he is interested in. In the case of a group of receptors or neurones in the brain, this can be done by recording a change either in the discharge in the afferent nerves from the receptors or in the efferent pathways from the neurones in the central nervous system; in the case of a stimulus to a secretory gland, this can be done by monitoring the secretion from the gland.

As well as showing that the stimulus is adequate for the purpose for which it is intended it must also be shown to be specific, that is, not to stimulate anything else. Thus, electrical stimulation of a group of neurones in the brain may also stimulate fibres which are running through the area. Similarly, the injection of a putative neurotransmitter into a nucleus in the brain may be adequate to stimulate the cells there but the transmitter may also leak away from the injection site, enter the cerebrospinal fluid and hence have more widespread effects. A drug given to affect the process of secretion of a gland at a cellular level might also alter the blood flow to the gland thereby changing the secretory activity of the gland by a secondary mechanism. Examples where a stimulus turns out to be less specific than the investigator intended are numerous.

If the investigator is aware of the problems, he can sometimes modify the design of his experiment accordingly. However, often he may have to settle for a stimulus which affects, for example, mainly one type of nerve fibre whilst not being able to guarantee that other fibres are unaffected. In other situations, a specific stimulus is just not possible; for example, you cannot stimulate only those unencapsulated endings in the atria which give rise to myelinated fibres if these endings are within a dense network of free nerve endings that give rise to unmyelinated fibres and that are also sensitive to the stimulus.

Another criticism often applied to a particular stimulus is that it is non-physiological. Obviously, if the stimulus involves electrical stimulation or the injection of fluid into the brain or the probing of a mechanoreceptor with a glass rod, such stimuli do not normally occur and hence, by one definition, are non-physiological. However, such stimuli should not be disregarded. Often physiologists use an artificial stimulus that can be carefully controlled and used reproducibly and compare the effects produced by this artificial stimulus with those found when more physiological but less easily controlled stimuli are used. In one series of experiments the effects of distending a balloon within the atrium (a 'non-physiological' stimulus), and of increasing the atrial pressure by changing the blood volume (the more physiological stimulus), on the discharge recorded from stretch receptors in the atrium were compared. In this example, the mechanical stimulus results in changes that are within the physiological range. In the sense that such stimuli mimic those found naturally or can produce results indistinguishable from natural stimuli, they have a claim to be called 'physiological'.

However, there are cases when commonly used stimuli do not mimic normal stimuli exactly. For example, maximal electrical stimulation of nerves tends to produce a synchronized discharge in all the fibres in a trunk and, if the intensity of the stimulus is reduced, then the larger fibres together with the fibres nearest to the electrodes will be stimulated. However, this may not apply naturally. For example, during a voluntary muscular contraction, it is the motor units supplied by the smaller diameter motor fibres that are activated first. It is not only the *order* in which fibres are activated that may differ during electrical and physiological stimulation; the *pattern* of discharge may be different under the two circumstances. These differences can be illustrated by comparing the ability of the β-adrenoreceptor-blocking drugs to modify changes in heart rate evoked by electrical stimulation of the sympathetic nerves with changes in heart rate of a similar magnitude which are reflexly evoked. The drugs are more effective in blocking the changes induced by electrical stimulation than they are in blocking the reflexly induced changes. This may be because the pattern of nervous impulses is different in the two cases. The synchronous increase in nervous activity during electrical stimulation will produce an increase in the release of transmitter from many of the fibres. In contrast, the reflexly induced change in nervous activity may produce an equivalent total release of transmitter but from only a few endings. Because of the competitive nature of the blockade the drug would be unable to block completely the effects of a large amount of transmitter released from only a few nerve endings. Similarly, the pattern of neuronal activity which is set up when (say) the motor cortex is electrically stimulated is unlikely to be identical to that when voluntary movement takes place. In such cases, the stimuli are 'non-physiological' but to what extent they produce misleading results is not known. Similar interpretive problems arise when the results of abnormally large stimuli (excessive heat load, for instance) are used. This problem is considered further in Chapter 8.

5.1.3 *Blockade*

Rather than stimulating a tissue or system, the alternative approach is to block the system thought to be involved in the response that is of interest. In terms of the design of the experiments, the problems that may arise are very similar to those already outlined.

A number of different techniques are available that can be used as blockers. Drugs are widely used experimentally but blockade, particularly of nervous pathways, can also be produced without drugs. Examples of other techniques are by cooling a nerve, by applying pressure to it or by driving an electrical current across it (and so hyperpolarizing it — anodal block). Regardless of the type of blockade employed, the experimenter must always consider its adequacy and specificity. It is important to check, for example, that the dose of a blocking agent is adequate to prevent the effects of an exogenously administered agonist. Similarly, if nerve blockade is being produced by cooling, pressure or anodal block, the experimenter must check that the nerve is not capable of transmitting action potentials along this portion.

As well as showing that the technique is capable of producing a blockade of the system under investigation, it is important to consider what other effects it might be having. For example, some drugs are partial agonists, that is, they may stimulate a receptor as well as blocking the effect of another chemical on that receptor. Thus, at different doses, drugs may have different effects. Other drugs may initially stimulate and then provide blockade so it may be necessary to wait for the blockade to take effect.

Many drugs affect more than one system; for example, drugs such as captopril that interfere with the generation of angiotensin may also interfere with the breakdown of kinins. Similarly, some antagonists of histamine also antagonise the actions of 5 HT (5-hydroxytryptamine). Other drugs may have additional actions unrelated to specific receptor interaction. The important point is that the experimenter should be aware of the problems associated with his choice of intervention. The perfect blocking agent (like the perfect stimulus) does not exist.

Again, as with the stimulus, the type of blockade used must be standardized, for example, by standardizing the drug dose or the temperature to which the nerve is cooled. It may also be relevant to consider whether the blockade produced resembles that which might occur in some physiological or clinical circumstances.

5.1.4 *Ablation*

The same general points arise when ablation is used as the intervention. Thus, the experimenter must produce evidence that the lesion — be it by excision of the tissue or destruction by heat, chemicals or electrical means — has removed the relevant system and yet has not interfered with anything else.

Technically, removal of, for example, the nerve supply to an organ should be easy in that it only involves sectioning of the nerves, but problems may arise in deciding which nerves should be sectioned. There are large individual variations in the paths taken by different nerves. This may lead to an incomplete denervation. A complete denervation of a tissue may therefore necessarily involve the sectioning of pathways mediating other responses. Similarly, lesions within an area in the central nervous system may, in addition to destroying the intended group of neurones, section fibres travelling through that area. When lesions are produced by the injection of toxins, care must be taken that the toxins are delivered specifically to the intended site and do not diffuse away and, for example, enter the blood or cerebrospinal fluid.

Thus, it may not always be possible to produce a total disruption of one pathway without directly effecting others. If there is a possibility that others have been disrupted then this should be reflected in the conclusions that the experimenter draws from the data. As with the other types of intervention, there is the difficulty that the selectivity (or lack of it) accompanying an ablation technique need not be the same as that which occurs naturally. Again, the extent to which 'non-physiological' results are produced by these techniques is unclear, but the interpretive problems are further considered in Chapter 8.

5.1.5 *Replacement*

Blockade or removal of a system may be done either reversibly or irreversibly. Data obtained from reversible experiments is easier to interpret since it is possible to obtain a second control period after removal of the intervention. If irreversible techniques are used, it can be argued that any observed effects resulted from factors such as fatigue of the tissue. A suitable control experiment becomes of fundamental importance in such cases and enables the results obtained from the control animals to be compared with those obtained in the experimental group of animals. Irreversible methods include the removal of organs or tissue, nerve section or drugs whose effects cannot be reversed. Reversible methods include the removal of a substance from the fluid perfusing an organ or bathing a cell, the blocking of nervous impulses by reversible blocking agents, local anaesthetics, cooling or anodal block or the inhibition of the action of a blocking drug.

Even if a technique is irreversible, useful information can be obtained by suitable replacements. For example, an experimenter showed that nephrectomy affected the baroreceptor reflex and he postulated that this was because the kidney could no longer produce renin. If he were then to replace the renin and this were to reverse the effect on the reflex, then this would provide a useful piece of supporting evidence for his hypothesis. Replacement of hormones and other chemicals after their production or release has been modified is a commonly used technique and has the added advantage that the concentration of added substances can be carefully controlled. Clearly, a modification of this technique would enable the effects of partial replacement (for example, different fractions of a cell homogenate) upon recovery of function to be investigated.

Transplantation is also used, in some studies, where one might be interested in divorcing the intrinsic and extrinsic control of an organ, or defining the source of a chemical such as a hormone. Here the organ is not removed but is transferred either to another part of the body or to a donor animal. Vascularization of the transplanted organ can occur spontaneously (in the case of some endocrines) or can be helped by connecting up the major vessels surgically. If the organ has been transplanted to its 'correct' site, then the nerve supply will eventually reinnervate the organ (at least partially). This is not the disadvantage it first appears to be, since it can be used as a 'natural' opportunity to investigate the effects of nerves.

5.2 **Measurement and recording**

Having designed the experimental protocol and chosen the intervention which is to be used, the investigator must also consider how he is going to measure and record the effects of his intervention.

5.2.1 *Documentation of results*

One of the first pieces of advice that we were given as students was to go and buy a hardback notebook for use in the laboratory. This was very sound advice (even though at the time it seemed too mundane to be of importance), and is so vital to successful research that it is advice that we now hand down to our own postgraduates! It is important to document carefully the progress of an experiment — whether successful or disastrous! Information jotted down in the course of an experiment — in addition to carefully annotated traces of the experiment — can be so useful when one comes back at a later date for further information, and may reduce the number of experiments that have to be carried out subsequently. Further, when experimental results are recorded, it is always easier to compare results or investigate anomalies if the format of recording is standardized; often photocopied results sheets can be a great help in this.

5.2.2 *Instrumentation*

Most scientific investigations will involve the use of instruments. The range of possible physiological variables which may be measured is enormous and consequently the number of instruments which a scientist might use is potentially huge. Any instrument which is purchased is normally accompanied by a manual from the manufacturers which gives many of the specifications of the machine and detailed instructions as to its use. The purpose of the following account is to give the reader some general advice as how to determine the characteristics of an instrument and to assess the suitability of a particular instrument for the measurement of a particular variable. Often, of course, such judgements ought to be made before purchasing an instrument.

Let us now consider those characteristics of an instrument that will most often concern us.

5.2.2.1 *Precision and accuracy.* According to both the dictionary definitions and common usage the terms accuracy and precision are used interchangeably. However, in the scientific literature these terms each have a very distinct meaning.

Precision of an instrument. The precision of an instrument is a measure of the variability or random error associated with repeated measurements of a sample, that is, the reproducibility of the measurement. Thus, a measure of the precision of an instrument can be obtained by making repeated measurements of a single sample and calculating the mean and standard deviation of the measurements. The smaller the observed standard deviation, the greater the precision associated with the measurement. The standard deviation is often expressed as a percentage of the mean (100 × SD/mean) and this gives rise to the expression 'with a precision of x per cent'. Expressed this way, precision is the same as the statistical term 'coefficient of variation'.

The absolute value for the precision of a measurement is not as important as the relationship between the precision of the instrument and the size of the change in the variable that you are measuring. For example, if the variable alters by 1 per cent this cannot be detected by an instrument having a precision of 10 per cent, but will be detected if the precision of the instrument is, say, 0.1 per cent. However, if the variable alters by, say, 50 per cent then an instrument which measures that variable to within 10 per cent is perfectly acceptable.

When the precision of an instrument is quoted by a manufacturer then, naturally, the optimum value is quoted. This value is often obtained under a particular set of conditions. For example, it is likely to be the precision with which measurements can be made over a particular section, often the mid-point, of the range of values that the instrument is capable of measuring. Measurements made at either end of the range can often be rather less precise than those in the middle of the range. Temperature may also influence the precision of an instrument as well as the substance being measured. For example, many blood gas analysers will measure a partial pressure of gas in air with a greater precision than the same value for the partial pressure of the same gas in blood. Sample size may also be important. In some cases, the development of instruments designed to measure a small sample (and necessary to measure samples obtained by micropuncture of kidney tubules, for instance) may result in a loss of precision of the instrument.

Thus, the experimenter needs to do rather more than merely look up the precision of the instrument in the manufacturer's specifications. What he must determine is the precision of the machine under the conditions of his particular experiments.

Accuracy of an instrument. The magnitude of the random error associated with a particular measurement is easy to determine. It is, however, of equal importance to know whether or not a systematic error is present. This is a measure of the accuracy of the instrument, that is, how reliably the instrument actually measures what it is supposed to. Systematic errors are often much more difficult to detect than random errors.

At the simplest level, how would you assess the precision and accuracy with which a ruler measures the length of a piece of string? The precision can easily be estimated by taking repeated measurements, but it could be that the ruler is capable of making very precise but completely inaccurate measurements. Comparing the values for the length of a piece of string measured by two different rulers could help — but what if both rulers had been manufactured using the same defective machine? In this particular example, the problem is easily solved as the length of the ruler can be compared with an internationally accepted (but quite arbitrary) standard of length. In general, the concept of a 'standard' against which machines can be measured is a useful one. A standard can be defined in a number of ways. Thus, it might be in physical terms (as above), in chemical terms (consider the definition of a molar solution), or in biological terms (producing a certain effect in a bioassay system). Having defined the standard, then the machine can be compared directly with this or with a sample that itself has been measured against the standard.

For example, if one wanted to evaluate the accuracy of a machine that is being used to measure the partial pressure of carbon dioxide in a liquid, such as blood, then the calculated value obtained when known concentrations of acid and sodium carbonate are mixed in a sealed container could be compared with the value obtained from the machine when a sample of the mixture was measured. Alternatively, the machine could be standardized against commercially available samples, the exact partial pressure of carbon dioxide in which was known. In effect, one is calibrating the machine against known standards (see below, p. 51). However, even though this is a necessary stage of the experiment — by which the apparatus is known to be measuring accurately — it will be performed under 'standardized' conditions that might not apply in the experiment itself. Thus, even though an electrode might measure pH accurately in a standard solution in a beaker, it need not do so, say, intracellularly. In such circumstances, a possible solution to the problem is to compare intracellular measurements made using that instrument with those made using a different technique. This means more than simply running duplicate samples through two machines of the same type since systematic errors may be a fault in the design of an instrument or probe. Measurements should be made using a number of different techniques and the results compared. In practice, if a measurement using the new instrument gives the same result as that using other established techniques, then it is assumed that neither technique has a systematic error attached to the measurement. It may, of course, also mean that both techniques have the same systematic error. If comparisons of many techniques result in the same result being obtained, then it becomes less likely that all the techniques have the same systematic error.

Just as the precision of an instrument may alter under different circumstances, so may the accuracy, and thus again it is important to assess the accuracy of a machine over the range of readings and under the conditions that are going to be employed in the investigation to be undertaken.

5.2.2.2 *Frequency response*.

The speed with which an instrument records a measurement of a variable is another important characteristic of a machine which must be determined before using the instrument in a scientific investigation.

If single measurements are to be made, for example, of the concentration of a substance, then it is important to know how quickly the instrument responds so that measurements can be made after steady-state values have been obtained. This is normally done by estimating what is known as the '95 per cent response time' of the instrument, that is, the time taken to obtain a readout which is 95 per cent of the final value. Estimates of the 95 per cent response time are made by injecting a sample into the machine and then noting the readout at frequent and regular intervals after injection. The values that the instrument gives are plotted against time (Figure 5.1). Using this method, the highest steady-state value can be measured and this is given a value of 100 per cent. The time taken for the instrument to record a value of 95 per cent of this maximum can then be readily estimated from the graph.

This technique also allows an estimate of the stability of the machine to be made. In some instruments, once the maximum value has been obtained, this value can be maintained for a considerable time. In other instruments, because of the instability of the machine, having once reached a steady-state value, this may then drift in either direction. If the drift is steady, it may be possible to correct readings by taking

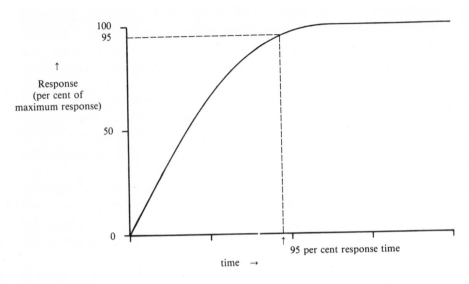

Fig. 5.1   Diagram to show how the 95 per cent response time of an instrument is calculated.

into account the passage of time. In any case, it is wise to continually recalibrate instruments throughout an experiment (see later).

Once the characteristics of the instrument are known, then the time at which measurements can be taken is established, thus reducing the random errors resulting from taking measurements at different times.

A knowledge of the frequency response of an instrument is even more important when one wants to obtain a continuous readout of a variable. In physiology, there are numerous occasions when we want to measure the pattern of changes over a period of time — for example, the changes in the composition of alveolar air during the respiratory cycle; in pressure in the chambers of the heart during a cardiac cycle; in the tension developed by a muscle during a sequence of contractions; or in potential when an action potential is generated in a neurone. Under these circumstances, if the frequency response of the instrument is inadequate, then the maximal and minimal values will not be recorded and therefore the instrument will produce a record which is said to be 'damped'. However, some degree of damping may be beneficial under certain circumstances as it allows a smoother record to be produced.

In practice, modern amplifiers rarely have a frequency response which is limiting, except when used in some neurophysiological studies. The limiting features of a set of apparatus are usually either at the sensor, for example, the cannula used to measure blood pressure, or at the recorder. Many pen recorders have dynamic characteristics which are inadequate even to reproduce faithfully the changes in arterial blood pressure during the cardiac cycle. For neurophysiological studies, the frequency response required to record action potentials is such that pen recorders are completely inadequate, and for faithful reproduction the changes in potential must be displayed on a recorder having a high frequency response, such as an oscilloscope.

The frequency response of a recording system is also particularly important if one is measuring the rate at which a value changes with time. For example, dP/dt_{max} (the maximum rate of change of pressure in the left ventricle) is used as an index of the inotropic state of the heart and, for this measurement to have any meaning, the frequency response of the recording system must be rapid enough. An example in which an even more rapid response is required is that when the rate of change of membrane potential during an action potential is being studied.

Details as to how to assess the frequency response of a recording system and how to determine if it is adequate for a particular purpose are outside the scope of this book.

5.2.2.3 *Calibration of instruments.*

At the start of each experiment, it is important to calibrate carefully all instruments to be used. Even though the actual techniques used will obviously vary, the principles of calibration remain the same.

The first step is to set up the instrument so that, when the value for the variable to be measured is zero, then the readout from the machine is zero. If it is not possible to adjust the readout, then the value of the readout when the variable was zero should be noted. The next step involves putting samples of known values into the machine and for each value recording the readout from the machine. A calibration curve can then be plotted (Figure 5.2) and this will enable the value of an unknown to be interpolated from the readout of the machine. Such curves are often non-linear and therefore estimates can only be made within the range that has been calibrated. Particular emphasis should be placed on the non-linear portions during calibration. Some instruments are designed so that the relationship between the readout and the value of the variable is linear, and thus only a two-point calibration is required. It is still wise to check the calibration of such machines against a range of known standards from time to time since this linearity may be lost under some conditions. It is essential to check the calibration of

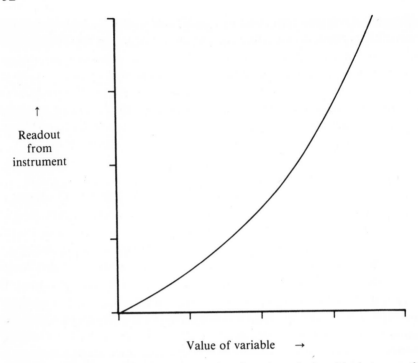

Value of variable   →

Fig. 5.2   Typical calibration curve showing the relationship between the readout from the instrument and the value of the variable being measured.

instruments at the start and finish of each experiment to check for any drift. In practice, this means that a machine may be calibrated several times during the course of a day.

Measurements obtained from any machine are only as reliable as the machine itself and it is vital that the experimenter knows precisely what the instrument is capable of doing. Results obtained from either an inappropriate instrument or one which has not been correctly used, calibrated and maintained are meaningless. The appearance of such data in the literature often misleads other scientists working in the field when they try and repeat others' experiments.

5.2.2.4 *The recording apparatus considered as a whole.* When assessing the characteristics of a recording setup, the experimenter must be sure to include all the different components in his tests. So often it is the small items of equipment made in laboratory workshops that alter the characteristics of a system. For example, the internal diameter, length and type of material used in making cannulae for the measurement of pressure will quite profoundly modify the frequency response of the recording system when taken as a whole; recordings of the potential difference across cells are likely to be more affected by the micro-electrodes used than the amplifiers and recordings of the changes in tension during a contraction will be dependent on the characteristics of the transducer used. In all cases, there is a need to compare results obtained from a particular setup with values obtained in other labora-tories, with other apparatus or from books, of 'normal' values.

5.2.3 *Chemical and biological assays*
The above comments concerning precision and accuracy refer to instru-ments which generally measure some physical characteristic of the system under consideration such as pressure, electrical potential difference, spectral emission, etc. Other types of measurement exist that are concerned more with the chemical or biological properties of the system. Obviously, all these areas overlap. Thus, even though it might be possible to assay the concentration of glucose in a solution by reference to one of its physical properties (its ability to rotate the plane of polarized light, for instance) one could also do so by a chemical property (its ability to react positively in Fehling's test) or a biological property (its ability to promote oxygen use, for example). In all cases, it is necessary to assess the precision and accuracy with which the measurements can be made and to calibrate the method by means of standard solutions.

There are two other areas where special care is needed to maintain the accuracy of the technique when chemical or biological assays are being performed.

5.2.3.1 *Storage.*
Unlike many physical measurements, those using chemical or biological assays are often made a considerable time after the experiment itself has taken place. Because of the lability of the samples of blood, urine or tissue, etc. consideration must be given to the circumstances under which the samples are stored. It is the respon-sibility of the experimenter to show that his storage procedures do not affect the assay. In practice, he will have performed a preliminary series of tests to establish that known concentrations of the substance in which he is interested do not change under the storage conditions used in his study.

5.2.3.2 *Specificity*. Biological fluids are rarely simple solutions! There is the possibility that a chemical or biological assay that is reliable when tested with solutions of the substance made up in distilled water is erratic when tested upon biological fluids. This could result from the presence of other substances in the test fluid that either react with the reagents or biological tissue being used in the assay (so leading to a spuriously high result), or interfere with the assay (so leading to a falsely low result). In other words, the problem of specificity has to be dealt with. In a chemical assay method for glucose, for example, does the test measure only glucose (and then all or only some isomers?); or does it measure all reducing sugars; or all those with a hydroxyl group at position 2, for instance? If a piece of muscle tissue is being used to assay the amount of, say, adrenaline, are there other molecules that produce a similar effect upon the tissue? This problem can be particularly important in some recently developed assays in which competitive binding is a crucial part. Consider, for example, the case of radio-immunoassay. What exactly does a particular assay measure? Does it measure only the substance in which one is interested, or is it affected by breakdown products of this substance, or is there interference (or 'cross-reactivity') from other molecules, often of very similar structure? Will the assay detect the substance if it is bound to, for example, a plasma protein?

In all cases, it is the responsibility of the experimenter to show that his assay detects all of the substance in which he is interested and none of the other substances that might be present. Preliminary tests are required, of course, and these should pay particular attention to substances that are likely to be present in his samples and that might affect the assay. In some cases, it is valuable to compare results obtained by different assay techniques. It is generally assumed (as with instruments) that getting the same results with two independent assays argues strongly for both of them.

5.2.4 *Standard recording procedures and artefacts of recording*
Just as (see Section 5.1) the stimulus parameters must be standardized in some experiments, so too must sites of recording in others. Thus, the sites for recording blood pressure plethysmographically, for recording brain activity during sleep electroencephalographically and for recording electrical activity of the heart from the electrocardiogram all have been standardized. In these circumstances, in which the measurements and apparatus have been long established as producing useful results, the

possibility of artefacts of recording is small (we assume that all apparatus has been calibrated correctly, of course). However, when newly developed apparatus is being used or a new type of recording is being made, the presence of artefacts is a problem against which the experimenter must be on guard. Thus, using the approaches outlined in the previous sections, the experimenter can normally select an instrument that is capable of recording the changes in which he is interested. However, all changes observed are not necessarily of any physiological significance.

5.2.4.1 *Movement artefacts.* One difficulty that may arise is that recordings may be made which turn out to be artefacts of movement. These are particularly common when neurophysiological techniques are employed, for example, when microelectrodes are used for recording. Movements produced during the different phases of respiration or as a result of pulsation of blood vessels may displace the recording electrode and thus produce a signal. An experienced experimenter looking at the raw data will be able to differentiate between movement artefacts and true neuronal activity by examining the magnitude and time-course of the two. However, if the record that is displayed is of integrated activity or rate of activity rather than of the raw data, then these movement artefacts may be counted. Further, even though it is possible to design equipment that will count only those changes in electrical potential which occur within specified voltage-limits and time-limits, many recording systems do not discriminate between changes of different origin — that is, due to 'real' electrical activity rather than electrical artefacts resulting from movement.

Another example where movement artefacts may be important is during labour when fetal heart rate is measured by means of electrocardiogram leads and the electrical signals are displayed as beats/min. Again without the raw data it is difficult to assess whether any displayed changes in 'heart rate' are real or are artefacts of uterine and fetal movement. An examination of the raw data will allow this question to be answered, since the electrical changes due to cardiac activity and those due to movement have characteristics that are distinguishable to an observer. Ideally, to eliminate these problems of interpretation, the raw data and calculated result should both be displayed continuously. At the very least, the experimenter needs to be able to display the raw data intermittently when he suspects that movement artefacts might be present.

These points are not just refinements of technique; they might significantly alter the inferences drawn from an experimental result. It may be that the intervention changes the rate of respiration or the heart rate and thus the number of movement artefacts recorded per unit time. Following a particular intervention, movements may cause an electrode to be displaced; thus the recordings taken during the experimental period may not be from the same site as those taken during the control period, and therefore the two recordings cannot be directly compared.

5.2.4.2 *Selectivity in recordings.* As with stimulation and ablation (discussed above, p. 40), another aspect of recording that should concern the investigator is whether the recording technique produces biased results. For example, extracellular recordings of the activity in peripheral nerves are often made by dividing fibres from a nerve trunk until a single active fibre remains on the electrodes. Using this technique, it is easier to record from large myelinated fibres rather than from the small unmyelinated ones. Thus the recordings may be biased towards describing the responses in the large fibres which may not be representative in their response of all the fibres. Similarly, it is easier to place microelectrodes inside large cells than small cells so again this will tend to result in recordings being made unrepresentatively frequently from the larger cells. A similar bias in recording frequency may occur in micropuncture studies on, for example, kidney tubules, blood vessels or the ducts of glands.

When making peripheral recordings from single fibres by teasing away nerve fibres, activity in the fibres is often assessed by listening to the electrical signal played through a loudspeaker. Thus, unless the experimenter takes active steps to avoid it, recordings will only be made from the population of fibres which is spontaneously active and not discharging at rest. Of course, it may well be the latter group that mediate the response that interests the experimenter. Additionally, there may be a tendency to concentrate on the strands of fibres which show a *rhythmic* discharge. For example, afferents travelling in the vagal nerves from the walls of the airways in the lungs have a very characteristic pattern of firing which is in phase with respiration, whereas the small unmyelinated C fibres in the vagal nerves commonly have a slower and more irregular discharge. Thus the experimenter may find himself unwittingly selecting for further study those 'interesting' fibres that numerically represent only a very small percentage of the total number of afferent fibres in the vagal nerves.

Because of these biases unwittingly introduced into recording, it is often found that a particular effect was not noted for many years simply because no one had designed experiments to look at it. Once an unbiased experiment was designed, the effect was seen. The moral, therefore, is 'seek and thou shalt find'.

5.2.5 *The relationship between technical and theoretical advances*

As the examples given so far in this chapter indicate, many important scientific advances have been preceded by the development of new techniques of recording. For example, the discovery that heat was produced when a muscle shortened was possible because of the development of a galvanometer of sufficient sensitivity, the use of patch clamping has allowed us to describe the characteristics of channels within cells, and the development of microelectrodes has allowed us to measure the concentration of different ions within a cell. As a rider to this, our understanding of some bodily functions is limited by the absence of suitable techniques of measurement. For example, we are still unable to see the channels in membranes through which substances may pass. Studies of mechanisms in the nervous system in man are particularly limited by the absence of suitable techniques; for example, we know very little about the mechanisms underlying man's higher functions because with the techniques available at present we can devise, but not perform, the experiments. This important limitation to scientific advance is considered further in Chapter 8.

Chapter 6

The form, presentation and treatment of results

Having made his experimental measurements, the experimenter now possesses many results. In what form does he express them, how does he display them to their best advantage and what calculations does he need to perform upon them? To a very large extent, the answers to these questions depend upon the hypothesis that is being tested.

6.1 The form in which results are expressed

There will sometimes be a need to perform some calculations upon the results before they can be used in the most useful way. For example, a piece of apparatus might give a readout of heart rate when the experimenter is interested in the interval between heart beats, or the apparatus might record muscle length at any moment when it is the velocity of contraction that is required. In the first case, the required result is the reciprocal of the readout, and in the second it is the change in length per unit time; in both cases, the conversion is very simple.

6.1.1 *Mathematical transformation*
In other cases, the result has to be transformed mathematically in accord with the hypothesis being tested or with theory. For example, if the relationship between two measurements x and y is believed to be a linear one, then the results can be used immediately (Figure 6.1 shows an example of this). However, if one were investigating the flow of fluid through blood vessels of different dimensions in circumstances in which Poiseuille's law applied, then theory would predict an inverse relationship between the laminar flow and the *fourth power* of the radius; accordingly, even though flow and vessel radius would have been measured, they would not be compared in exactly this form.

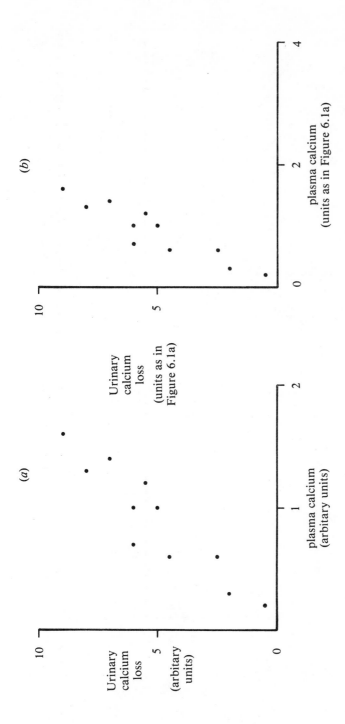

Fig. 6.1   The relationship between urinary calcium loss and plasma calcium; the same data are displayed in both Figure 6.1(a) and 6.1(b) but in 6.1(b) the abscissa covers twice the range.

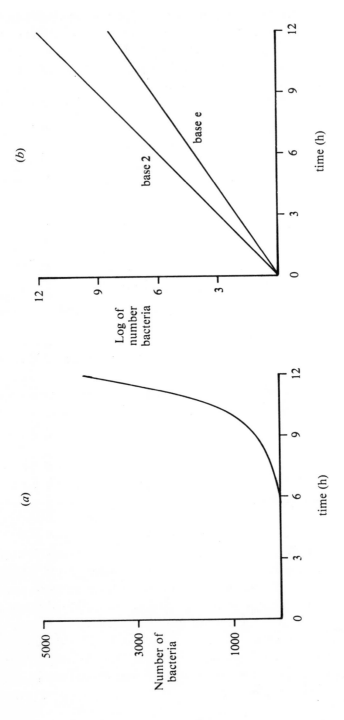

Fig. 6.2 The relationship between (*a*) the number of bacteria and, (*b*) the log of the number of bacteria and time; the bacteria are derived from a single bacterium at zero time and can divide hourly.

Another mathematical transformation is when a process is described exponentially. This could be the increase in number of cells when they divide (Figure 6.2), the decrease in concentration of a substance being lost from the plasma by filtration at the kidneys, or the decrease in concentration of an isotope by radioactive decay. In these cases the relationship between the two variables (time and number of cells or time and concentration) can be described by the equation

$$y = a \cdot e^{b.x}$$

(where a and b are constants). This can be transformed by taking natural logarithms into:

$$ln\ y = ln\ a + b \cdot x,$$

which has the form of a straight line ($y = c + m \cdot x$).

In other cases, the relationship between two variables might be a power function expressed as:

$$y = a \cdot x^b$$

(where a and b are constants).

Examples would be the relationship between body weight (x) and metabolic rate (y), or the relationship between firing frequency in sensory nerves (y) and stimulus strength (x). In these cases, logarithmic transformations would give

$$ln\ y = ln\ a + b \cdot ln\ x$$

from which it can be seen that the gradient of the straight line so produced is the value of the power b (note that the logarithm base in this case is unimportant).

6.1.2 *The units in which results are expressed*

Sometimes results are modified in other ways to render results between different animals, tissues or cells more easy to compare. Thus, a comparison between an infant and an adult is more meaningful if some account of their difference in size is taken. Similarly, the blood supplies to different organs might be a more useful measure of their need if differences in organ weight were considered also. This concept of expressing a result in terms of unit weight, whether the weight refers to the whole body, a whole organ, a tissue or a single cell can be very useful.

Consider the following examples:

(1) To express a dietary requirement in terms of 'per kg body weight'. This is particularly useful if the dietary factor is used by the whole body; for a factor like an anaesthetic gas which tends to be particularly soluble in liquid, whole body weight might not be as useful as some measure of the amount of adipose tissue in the body.

(2) To express the secretory activity of a gland in terms of 'ml of secretion per 100g tissue'. This gives a measure of the activity of a standard amount of glandular tissue.

(3) To express blood flow in terms of ml of flow per min per 100g of tissue allows us to determine whether the blood flow to a tissue is high merely because the tissue represents a large mass.

However, to express result in terms of 'per unit mass' as above might not always be the most appropriate way. Thus, in example (2) above, the measurement need not reflect accurately the activity of the whole gland, since it does not take into account the total weight of secretory tissue. If the *total* secretory activity is required, then expressing secretion in terms of 'per gland' rather than 'per 100g' is preferable. In a similar way, urea excretion in the urine might be expressed not only as 'g/h' or 'g/h/kg body weight', but also as:

'g/h/kg lean body weight' (to take into account its origin from tissues other than adipose tissue),
or as 'g/h/kg liver weight' (to stress more the role of the liver in manufacturing urea).

These examples illustrate the comment made earlier that the means of expressing results can depend upon the aim of the experiment.

Other examples where 'per kg body weight' would not be the best units are:

(1) To express heat production or oxygen metabolism in terms of 'per m^2 surface area'. (This takes into account the fact that metabolic activity and heat production are to maintain body temperature and heat is lost via the skin.) Sometimes this is expressed indirectly by considering the result in terms of $W^{2/3}$ where W is body weight. For a sphere, $W^{2/3}$ equals surface area, of course.

(2) To express protein content of a tissue in the units 'mg protein per μg of DNA'. (In this case, the assumption is made that the

number of cells in a tissue can be estimated from the number of nuclei it contains, each of which contains a constant weight of nuclear material.)

Closely related to this second example is the concept that an increase in the number of cells — hyperplasia — can be assessed by estimating 'DNA/g wet tissue', and an increase in individual cell size — hypertrophy — by estimating 'protein/μg DNA' or 'tissue/μg DNA'.

6.2 **Presentation of results**

Having decided upon the way in which the results are to be expressed, it is now important to display them clearly. Again, the best manner often depends upon the hypothesis being tested.

First, a decision has to be taken as to whether individual or pooled results are to be shown. During the course of the discussion, it will become apparent also that there is a link between the way in which the results are presented and the statistical analyses that will need to be performed upon them. Furthermore, a visual presentation of results can highlight some interpretive problems. We will illustrate these points by considering a number of different hypotheses.

6.2.1 *The relationship between two variables in a single subject*
If the hypothesis is that urinary calcium excretion is determined by plasma calcium concentration, then simultaneous measurements of both variables should be displayed and some means of quantifying any relationship between them will be required (Figure 6.1*a*).

When the correlation between two variables is sought, then a diagram such as Figure 6.1*a*, in which each point represents the results from a single sample of urine and a sample of blood taken at the mid-point of urine collection, is a useful means of display.

Some care is required with the scales of the axes. Thus the data of Figure 6.1*a* are presented again in Figure 6.1*b* but the abscissa now covers twice the range. This has no effect on the outcome of statistical tests or on the constants in the linear regression equation, but it does affect the clarity of presentation, the results in Figure 6.1*b* being unnecessarily cramped towards the ordinate. The scales should be chosen so that results cluster about a line drawn at 45° to the axes.

The choice of scales is important too in those cases when one or both of the variables exhibit a rather wide range of values. Here linear scales are inconvenient (either they make the graph too big or the scales have to use very large units). Examples would be if the decreasing threshold to light of the eye during dark adaptation were being considered (in which sensitivity can change by a factor of 100 000) or if some form of exponential growth were present. Thus, Figure 6.2 shows the number of bacteria derived from a single bacterium during the course of a 12 h period if each bacterium could divide every hour (see Figure 6.2*a*). Clearly, to accommodate the last value (2^{12}), the scale has to be such that most other values are squashed far too close to the abscissa. A logarithmic scale will help here (Figure 6.2*b*). The logarithmic base that is used is irrelevant to the shape of the curve (it will always be a straight line) and only alters the gradient. Again, a gradient of about 45° to the axes is clearest. In Figure 6.2*b*, both natural logarithms and logarithms to the base 2 have been used.

A logarithmic scale is also useful when ratios are considered. Suppose we have two variables X and Y both of which can independently take any whole number value from 1 to 100 with equal probability. What happens if we look at the ratio of X to Y? We might expect that the mean value of all ratios will be unity; this is not so. Thus, even though an equal number of values will be above and below unity — and so the median value will be close to our expected value — nearly half of the values will lie below unity, whereas the remaining half will range from 1 to 100. We will have a very asymmetric distribution (skewed to the right) and the mean will be considerably greater than unity. If, however, the log of each ratio is taken, then the mean of the logs will be zero (which is the log of 1) and the individual values will be normally or symmetrically distributed about this value. This has considerable statistical implications — since one is 'normalizing' the data by logarithmic transformation — as will be explained in Chapter 7.

Other instances where one or both of the scales are logarithmic are when exponential or power functions are involved and mathematical transformations have been performed. These have already been covered in Section 6.1.1.

6.2.2 *Display of data from more than one subject or animal*
6.2.2.1 *Correlation between two variables.* If results from more than one subject or animal are being used, then problems can arise. Figure 6.3*a* shows three subjects, each of whom shows a positive correlation, and Figure 6.3*b* another three subjects, none of whom individually shows a significant correlation. However, if, in each case, all data

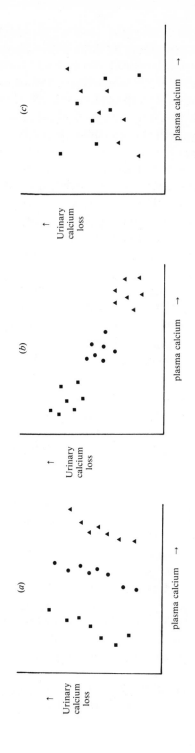

Fig. 6.3 The relationship between urinary calcium loss and plasma calcium. (*a*) Data from three individuals; in each individual, there is a positive correlation but when the data are pooled, there is no correlation. (*b*) Data from three individuals; in each individual, there is no correlation but when the data are pooled, there is a negative correlation. (*c*) Data from two individuals, one showing a negative correlation (■) and the other a positive correlation (▲) when the data are pooled there is no correlation.

points are considered, then a different result is obtained. In the data of Figure 6.3*a*, there would be no correlation overall, and in the case of the results of Figure 6.3*b*, a negative correlation would result. In Figure 6.3*c* is shown another example where there is no correlation when the pooled data are considered, even though each of the two subjects shows a correlation when treated individually, one negative and the other positive. In all these cases, the result from the pooled data does not describe the individuals satisfactorily. One of the assumptions made when correlation is being investigated is that the points are independent of each other. In the cases of Figure 6.3*a* and 6.3*b*, many of the data are certainly not independent but rather divide into three groups; with the data of Figure 6.3*c*, a similar criticism applies with the data being divisible into two related groups. In practice, it is unwise to mix a group of results from one individual with groups from others. One approach is to estimate whether, for each individual in turn, a correlation exists. However, if each animal or subject can give only a *single* pair of values, the presence of a correlation between these two variables can be assessed from data obtained from a group of individuals (where *n* equals the number of individuals). In this case, of course, the assumption must be made that all individuals in the sample have been randomly and independently drawn from the same population.

6.2.2.2 *Data from experiments with a paired design.* Another occasion when individual results are informative is in a paired experimental design when the same material is being used for both control and experimental phases of the experiment. For instance, if one were testing the hypothesis that the respiratory minute volume at rest is greater during wakefulness than during sleep, then some indication of this comparison would be required (Figure 6.4*a*). The results shown in Figure 6.4*a* illustrate an experiment in which the same animal was tested twice, once asleep and again awake (though, as has already been described in Chapter 3, the order in which these parts were performed should have been randomized between different animals). Such protocols will enable the statistical assessments to concentrate upon the *difference* between the two values rather than their absolute values. The distances along the (broken) abscissa between the two values are arbitrary and for clarity only. Such results are sometimes presented in a different form:

(1) one of the two values is set at some arbitrary value (generally zero) and differences from this are shown (see Figure 6.4*b*);

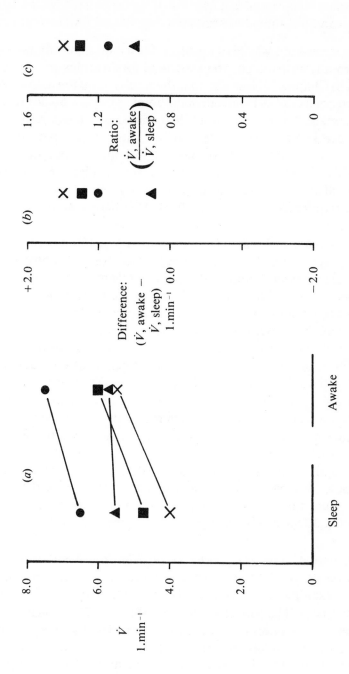

Fig. 6.4 Ventilation, $\dot{V}$ during sleep and when awake in four individuals expressed as (a) absolute values awake and asleep; (b) differences between $\dot{V}$ awake and asleep; (c) ratio of $\dot{V}$ awake to $\dot{V}$ asleep.

(2) one value is expressed as a fraction of the other (see Figure 6.4c); in this example, unity has been used, but 100 per cent is common.

When expressed this way (as a ratio), the distribution of results need not be normally distributed; the statistical implications of this are considered in Chapter 7. Note also that the results of Figure 6.4b could be expressed not as values from individual animals but as a pooled result from all animals. The results would then appear as a mean value for the difference in ventilation rate together with its standard error. Statistically, one would wish to know if this mean value differed significantly from zero. The results as expressed in Figure 6.4c could also be pooled but, as has already been described, since they are ratios they might require logarithmic transformation before doing so.

6.2.2.3 *Data from experiments with an unpaired design.* In other studies, different animals may have been used for the control and experimental groups in an unpaired experimental design. For example, the experiment might have been designed to see if perfusion of the ventricles of the brain with an artificial cerebrospinal fluid (c.s.f.) that is low in glucose increases food intake. Here a comparison is required of the food intake of two groups of conscious animals allowed free access to food, one group being perfused with c.s.f. of normal composition and the other with glucose-deficient c.s.f. (Figure 6.5).

In Figure 6.5, the results from five groups of animals receiving different treatments are given; for each group the mean food intake and its standard error are shown. In this respect, therefore, the presentation differs from Figures 6.1, 6.3 and 6.4 in that a pooled result, not those from individuals, is presented. The ordinate scale starts at zero, but need not. However, if the scale does not start at zero, then the differences (and standard error) appear to be magnified and this can be misleading as far as visual inspection is concerned.

6.2.2.4 *Display of data as a frequency distribution.* Another means of pooling results is to display them in some form of frequency distribution, one example of which is a frequency histogram.

The data that will be considered are in Table 6.1. The 33 values are arranged in ascending order and are to be considered as having been measured on some equal-interval scale (that is, they are of interval status, see Section 7.1.2.1), such as 'cm', 's', 'g.cm^{-3}', 'mU activity per g tissue', etc. They have also been rounded off to the nearest whole

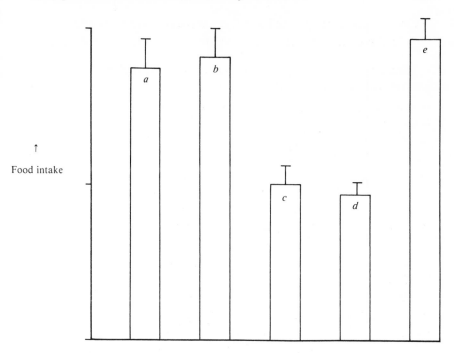

Fig. 6.5   Mean ± SE of food intake in different groups of animals: (*a*) un-operated animals; (*b*) sham-operated controls; (*c*) operated cannulated, un-perfused controls; (*d*) operated, cannulated, perfused with artificial c.s.f. containing normal glucose concentration; (*e*) operated, cannulated, perfused with artificial c.s.f. containing low glucose concentration. Food intake in group (*c*) is significantly lower $P < 0.05$, than in group (*b*) and in group (*e*) is significantly higher than in group (*d*), $P < 0.05$.

number for convenience. The results have also been plotted as two frequency histograms (Figure 6.6*a*, 6.6*b*); these histograms differ in the choice of limits for the class intervals, but not in the width of these intervals. In both cases, the intervals chosen for collection of the data were arbitrary. The data in Fig. 6.6*b* appear more skewed and there is even the suggestion of a smaller, second peak.

The results could also be displayed as a cumulative frequency histogram or ogive (Figure 6.6*c*). This can be a very convenient way of comparing two distributions visually and it is the basis of the Kolmogorov–Smirnoff test (see Chapter 7) which specifically investigates if two distributions differ significantly. A further type of histogram frequently used in neurophysiological studies is an interval histogram. In this the intervals between successive action potentials are measured

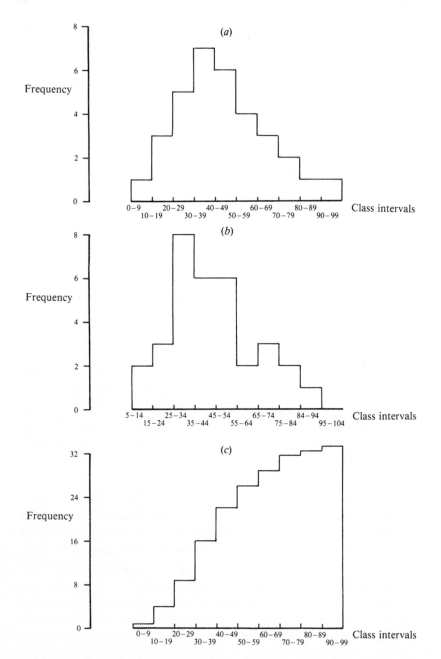

Fig. 6.6   Data from Table 6.1 displayed as frequency histograms (*a* and *b*) in which the limits for the class intervals differ and as a cumulative frequency histogram (*c*).

Table 6.1 Tabulation of data for frequency distributions

8	42
12	43
17	43
19	46
23	47
25	49
26	52
26	52
28	52
31	58
33	60
33	65
34	67
35	72
37	78
38	83
	91

and the results are expressed as a histogram. The class intervals will be of the form '0–10 ms', '10–20 ms', etc. and the ordinate will show the frequency with which such intervals between spikes were found in a span of data. Such histograms are often skewed to the right, particularly if the data show bursts of activity separated by comparative quiescence.

6.2.3 *Describing the time-course of an effect*

An experimenter may be interested in the relative time-courses of different effects. For example, if he is testing the hypothesis that local warming of the skin produces an increase in sweating *before* deep body temperature has been affected, then the time-courses of local warming, deep body temperature and sweating must be displayed simultaneously (Figure 6.7). It is most convenient to present the results so that the time axis is common to all traces. When presented this way, it is fairly easy to compare visually the changes in cutaneous heating, cutaneous sweating and deep body temperature. It will be noticed (see Figure 6.7) that the sweating rate was stable both before the stimulus began and before it ended, important points as far as the protocol and interpretation of the results are concerned. Further, the changes in deep body temperature clearly lagged behind those of sweating indicating that deep body temperature could not have *initiated* the changes in sweating.

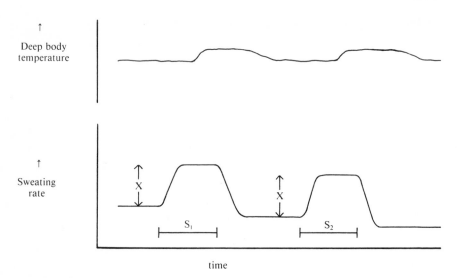

time

Fig. 6.7   Changes in deep body temperature and sweating rate occurring in response to local warming of the skin as indicated by the bars S_1 and S_2. X represents the absolute change in sweating rate in response to local warming and in both tests the value for X is the same.

Again, the presentation of individual rather than pooled results is desirable. With pooled results, it is only the average relationship between the two time-courses that is shown and an occasional anomalous result (that is, a change in deep body temperature *before* that of sweating rate) — a result that refutes the hypothesis — would not be discerned. It is worth noting that, as will be described in Chapter 8, it is often the recognition of a contradictory result which gives an opportunity to devise a more appropriate hypothesis.

6.3   Usefulness of visual presentation of data

Such presentations of data in the form of graphs, histograms or traces serve a number of functions. The first is that they readily enable the experimenter to see if the results obtained are at least approximately as predicted by the hypothesis. Even at this stage, a change in hypothesis might be indicated. Consider the results in Figure 6.8 that were obtained in an experiment to test a hypothesis predicting that X and Y are positively correlated. Had these results been subjected to a linear regression analysis, almost certainly a significant positive correlation between X and Y would have been found, and values for the gradient of the

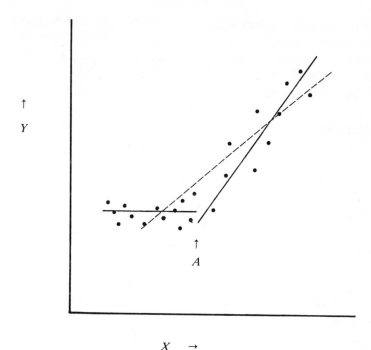

Fig. 6.8 Relationship between X and Y. Dotted line represents the calculated regression line for all the data. However, visual inspection of the data shows a discontinuity at A and the unbroken lines have been fitted by eye to the values when X is either greater than A or less than A.

best-fitting line and its intercept upon the Y axis could have been calculated (inserted as dashed line). However, inspection of the raw data of Figure 6.8 indicates that a single straight line (the *assumption* behind the linear regression analysis) is inappropriate; instead, there seems to be a discontinuity in the results at A. Two possible solutions to this problem exist. First, it might be appropriate to divide the results into two parts and analyse each separately (see inserted full lines). If this course of action were taken, then the original hypothesis would need modification in some way. It may mean that there is some threshold value of X below which Y is independent of X. The decision as to where to divide the data might be based upon visual inspection of them, but this is unsatisfactory because it is arbitrary. Much better would be to postulate some biological mechanism to account for such a discontinuity and this might be an example where a 'wrong' result gives rise to a better hypothesis. To take a concrete example, there is now some evidence that various drugs which modify baroreceptors have a more pronounced effect on the reflex response to a fall in blood pressure than to a rise in blood pressure. Thus, if an experimental protocol had been adopted in which resting blood pressure was always taken as the control value and it was then either raised or lowered to investigate the reflex response, it would be entirely appropriate to consider the possible modulation of the response to falls in blood pressure separately from the response to rises. A second solution is to devise another hypothesis which would be biologically plausible and which, mathematically, would predict a non-linear but continuous relationship.

In other experiments when one is looking at the effect of a particular intervention on the correlation between two variables, it may be found that the intervention affects one portion of the relationship more than another, for example, either the high or the low values of a variable. Again, an inspection of the data before carrying out a linear regression analysis will reveal this and it may then be appropriate to deal with the data in two or more groups and to modify the original hypothesis accordingly.

The importance of a visual presentation of the data can also be seen by considering another example where the experiment has been designed to assess the importance of neural factors in the control of muscle blood flow during exercise. A display of results from such an experiment — that is, of blood flow during rest and exercise, with and without intact nerves — is given in Figure 6.9. It can be seen from this figure that it is the *time-course* taken to attain stable values during exercise that is changed after denervation. This implies that an investigation of the effects of nerves upon the response to exercise might require more emphasis to be placed upon rates of change than was given in the original protocol. That is, not only would the original hypothesis be changed — the nerves exert an important effect upon the *rate of change* of the blood supply to an exercising muscle — but also the means of testing it (by making more frequent measurements, especially at the onset of exercise) would need modification.

Again, as this example indicates, it is important to choose the scales of both axes appropriately (see Section 6.2.1). More detail can often be seen by expanding the scales, but there is a limit. This limit is reached when further expansion would over-emphasize variations which result from the lack of precision of the apparatus involved. There is an analogy here with the process of photographic enlargement; after a while, further magnification does not show more detail because the limit of resolution

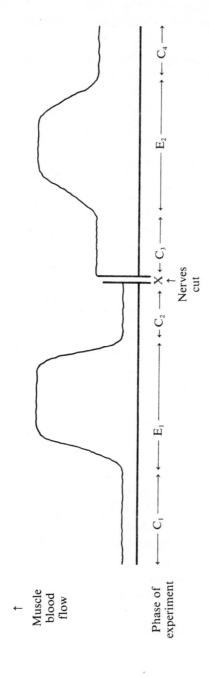

Fig. 6.9   The effect of exercise on muscle blood flow before and after nerve section (X). Phases C_1 and C_2, control phases with nerves intact; E_1, exercise phase with nerves intact; C_3 and C_4, control phases with nerves cut; E_2, exercise phase with nerves cut.

has been reached and all that results is an increase in display of the grain of the emulsion. With scientific experiments, too high a magnification by the recording apparatus results in the display of too much 'background noise'. This concept is expressed sometimes in terms of the 'signal/noise ratio'.

When assessing the change in blood flow during exercise, it is important to know the control or 'baseline' values — the blood flow during rest — above which the exercise values rise. Figure 6.9 also indicates that these change after nerve section (X) but that they are equal before and after each exercise period (that is, the values during phases C_1 and C_2 are equal as are those during phases C_3 and C_4). Such results offer clues as to the control of blood flow at rest, since they suggest that the nerves normally produced a tonic inhibition of blood flow. These differences in the control values are also important when calculating the increase in blood flow during exercise in the two groups. Some of the problems associated with calculating the difference between the control and experimental values will be considered in the following section.

6.4 Using the results to make calculations

As a general rule, the calculations are determined by the hypothesis that is being tested. However, a number of problems can arise when individual cases are considered in detail.

6.4.1 *The control values*
Consider the results as illustrated in Figure 6.7. As the experiment progressed, the 'baseline' or control value for cutaneous sweating became lower. Whatever is the explanation of such a result, the problem is to determine the most appropriate baseline against which the result in each experimental phase is to be compared. Obviously, the value in the pre-experimental control phase is acceptable but this might underestimate the response to stimulation as indicated by the lower post-experimental control values of sweating. An alternative is in some way to take into account both pre-experimental and post-experimental controls but the relative weighting of these is necessarily arbitrary. The simplest approach is to compare the experimental result with the mean of the pre-experimental and post-experimental controls. More importantly, whatever is decided, this same weighting must be made each time.

There are occasions when changes in baseline values are to be expected and even give useful information about the physiological processes involved. Thus, after drug addition or the sectioning of nerves, changes might indicate effects due to surgery or the removal of pre-existing tonic activity. Consider, for example, the effects of nerve section at X in Figure 6.9. In this case, the rise in muscle blood flow produced by section of the nerves might represent some form of 'stress' (? adrenaline release) or — as was suggested above — a removal of the tonic inhibition of blood flow by sympathetic nerves. Since the effect was not transient (the blood flows during the control phases C_3 and C_4 were equal), this might be taken to argue against the cause being 'stress' as this would be expected to die away. However, this point might be clarified if further control experiments were carried out; experiments with reversible nerve block, with nerve section but not exercise, and with 'sham' operations might all play a part in this.

This problem of a changing baseline is quite common in physiological research. Let us consider an example to illustrate this point. It has been observed that blockers of the renin−angiotensin system reduce the blood pressure response to carotid occlusion. However, these drugs also reduce blood pressure in these experiments. Since, in many biological circumstances, the relationship between the stimulus and the response is non-linear, changing the baseline values during the control phase will therefore be predicted to change the size of response. We therefore have the problem of deciding whether the drugs are actually modifying the response to carotid occlusion or whether the change in the response is simply secondary to the change in the baseline.

There are two possible ways of tackling this problem during the experiment itself. Firstly, one could change the blood pressure over a range and see if this modifies the responses to carotid occlusion. Alternatively, one might design the experiment so that the baseline does not change after giving the blocking drug, perhaps by giving yet another drug which counteracts its effects. In this example, it would be necessary to infuse a hypertensive agent so that blood pressure did not fall when the blocker was given. (This, of course, may pose further problems of interpretation since this second drug may itself be modifying the response to carotid occlusion or interacting with the first drug as well as affecting the baseline values of blood pressure!)

However, an experimenter will often not try and correct the changes in baseline (or be unable to); in such cases, care is needed when assessing the size of an experimentally mediated change (see below, p. 79).

A consideration of these points will indicate that there may be different 'controls' for different parts of an experiment. For each part, the controls can be compared with textbook values or values from other laboratories. This concept that more than one 'control' can exist is illustrated again, but in a more elaborate manner, when the results of Figure 6.5 are considered. This experiment was designed to test if lowered glucose concentrations in cerebrospinal fluid (c.s.f.) would increase food intake. Clearly, such an experiment required the animals to be conscious and to have free access to food. The glucose content of the c.s.f. was modified by perfusion of the ventricles with an artificial c.s.f. and this required prior surgery (under anaesthesia) to implant input and output perfusion cannulae. Since preparations for the experiment are fairly elaborate, a number of control and experimental groups is required. These are:

(1) unoperated animals ('unoperated' controls);
(2) animals that hav been anaesthetized and have been operated upon to prepare for implantation of the cannulae *except that* the cannulae have not been implanted ('sham-operated' controls);
(3) animals that have had cannulae inserted but there has been no subsequent perfusion through them ('unperfused' controls);
(4) animals with perfusion through the cannulae of an artificial c.s.f. of normal glucose concentration ('normal c.s.f.' control);
(5) animals with perfusion of a glucose-deficient, artificial c.s.f. (the experimental group).

It will be noticed that items (3) to (5) can be combined in the same animals, enabling paired comparisons to be made. Notice also that the different controls are required here to ascertain the effects of surgery, cannulation, etc., whereas in Figure 6.9 (considered above), there was the additional possibility of tonic activity in the nerve that was being cut (or, in more general terms, in the structure that was being ablated).

Comparisons between *adjacent* columns in Figure 6.5 enable certain inferences to be drawn unambiguously since the protocols differ in only one aspect; comparison of non-adjacent columns leaves doubt as to the cause of any differences. For example, comparison of columns (*c*) and (*e*) investigates the effects of perfusion of the cannulae with artificial c.s.f. *and/or* the effects of low glucose concentrations in c.s.f. The main aim of the experiment is considered in columns (*d*) and (*e*); here the significant difference indicates an effect of low glucose *alone*. However, these experiments do not identify whether it is the metabolic or osmotic effect that produces the result and a resolution of this problem would require further control experiments in which other substances with the same osmotic pressure as the glucose were added to the c.s.f. Since columns (*c*) and (*d*) do not differ significantly, then it can be concluded that the artificial c.s.f. is not causing artefacts (either by virtue of its composition or the perfusion pressures involved). By contrast, the significant difference between columns (*b*) and (*c*) suggests that implantation of the cannulae is decreasing food intake, even though the general operation procedures do not have an effect upon food intake (columns (*a*) and (*b*) do not differ significantly). As a result of these findings, it would be useful to investigate if low glucose promoted food intake in animals implanted with different types of cannulae (or implanted with cannulae at different brain sites) and in whom food intake had not been depressed. Note also that comparisons of the results from groups (*d*) and (*e*) with those from group (*a*) acting as a control would have led to the mistaken conclusions:

(1) that perfusion with artificial c.s.f. decreased food intake; and
(2) that perfusion with an artificial c.s.f. low in glucose had very little effect upon food intake.

6.4.2 *Calculating experimental changes*

Consider, for instance, Figure 6.7 and particularly the increase in sweating produced by cutaneous heating. The problem associated with a changing baseline has been mentioned already. Suppose further that it was required to know if the two cutaneous stimuli, indicated in Figure 6.7 by S_1 and S_2 exerted a different effect upon sweating rate. If the increase is defined as the absolute increase above control values, then the effects of stimuli 1 and 2 are identical (both equal X); if it is defined as the final plateau value, then stimulus 1 has a greater effect; and if as the proportional increase above the control value, then stimulus 2 has a greater effect. Similar considerations apply when assessing the effect of drugs (which often change baseline or control values) or when assessing the results of Figure 6.9 in which the final plateau values (phases E_1, E_2) are the same but the increase above control values is less (both in absolute size and as a proportion of the relevant control value) after nerve section.

Which of these possibilities is used might be determined by convention or by the hypothesis that is being tested. For example (referring again to the results of Figure 6.7), if the hypothesis were that cutaneous stimulation switched the skin into a 'sweating mode' in an all-or-none manner, then the final plateau values would be predicted to be the same; if the hypothesis were that cutaneous stimulation produced an effect independent of other factors, then the observed increase, X, would be predicted to be constant; and if it were suggested that cutaneous stimulation produced an effect via mechanisms serving other functions also, then the size of the effect might be expected to depend in some way upon the baseline values during the control phases.

Another example is shown in Figure 6.10a in which the effects of two gases, A and B, upon the respiratory minute volume are considered. A and B both produce significant increases in ventilation over control values, and A and B together produce the largest change. Suppose further that this change is significantly different from the effects of either A or B separately. A simple hypothesis would be to postulate that the effects of A and B together were additive; alternative hypotheses might be that the effects of A and B together were more than the sum of their individual effects or that the effects of A and B together were less than this sum. (Such hypotheses are often postulated when dealing with the effects of drugs and they invoke the concepts of synergism and antagonism respectively.) A distinction between these hypotheses cannot be made by showing that the effects of A and B together are different from the effects of either alone. That is, the statistical tests performed upon the results of Figure 6.10a are inappropriate to enable one to distinguish between the rival hypotheses.

A more appropriate analysis would be to investigate the relationship between the predicted result if the effects of giving A and B separately are summed and the observed result when A and B are given together (Figure 6.10b). A linear regression analysis would then establish the gradient of the line describing this relationship. If this gradient differed significantly from unity, then the null hypothesis — that there was no difference between the effects of A and B given together and the sum of the effects of A and B given separately — could be rejected. Which alternative hypothesis to accept could be obtained by inspection of the results (Figure 6.10b) and a knowledge of the value for the gradient of this line.

In summary, therefore, visual presentation of results often helps the experimenter decide upon the calculations that will test the experimental hypothesis. (Visual presentation might also become an important illustration of the scientific argument when the results are submitted for publication.) But the hypothesis can be formally tested only by a statistical evaluation of the results, and it is to this that we now turn.

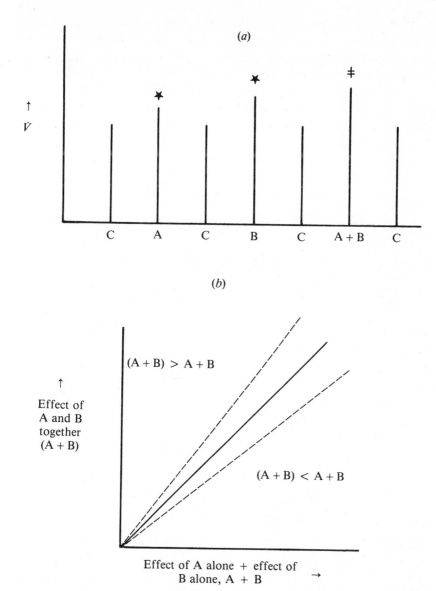

Fig. 6.10 (*a*) Effect on ventilation, $\dot{V}$, of breathing various gas mixtures; C = control, A = gas A and B = gas B. *, $P < 0.05$ compared with controls; ‡, $P < 0.05$ compared with A or B alone. (*b*) Relationship between effect of A and B together (A + B), and the sum of the individual effects of A and B, A + B. The unbroken line represents the predicted result if (A + B) = A + B. The broken lines represent the 95 per cent confidence limits about this line.

Chapter 7

The Statistical Assessment of Results

Whenever a series of experiments is performed, a range of results is obtained. This variability is partly 'experimental error' and can be attributed both to the apparatus being used for measurement and to the experimenter himself. These aspects have been discussed in earlier sections and we repeat the need for meticulous care in the maintenance, calibration and use of any measuring apparatus so that such sources of error are minimized. However, in biological systems a further source of variability — biological variation — exists. As has been described in Chapter 3, the assignment of individuals to different groups should be determined randomly. In this way, interindividual differences should be 'averaged out' and the bias between the groups should be minimal. Even so, with some results, the intragroup variation might be such that it obscures differences between them. Statistical tests enable both intergroup and intragroup variation between and within groups to be compared so that a decision can be taken as to whether observed differences could have arisen by chance or resulted instead from a real difference between the groups.

The use of statistical tests necessarily involves a large amount of repetitive calculation, and this rarely endears itself to the student! With the widespread availability of calculators and computers this chore has been circumvented, but another problem remains, namely, the choice of statistical test that is most appropriate for a particular set of results.

The present chapter is not intended to be a detailed description of different statistical tests (for which standard textbooks should be consulted). Its aim is to explain how the most appropriate statistical test is determined by the hypothesis that is being tested, the general experimental design and type of measurement that has been made.

Some idea of the types of tests that are available is given in Table 7.1.

Table 7.1 Summary of types of statistical test

Level of measurement	One-sample case	Comparing two samples		Comparing more than two samples		Measure of correlation
		Paired samples	Unpaired samples	Paired samples	Unpaired samples	
Nominal	1. Binomial test 2. χ^2 one-sample test	McNemar test for the significance of changes	1. Fisher exact probability test 2. χ^2 test for two independent samples	Cochran Q test	χ^2 test for k independent samples	Contingency coefficient: C
Ordinal	Kolmogorov–Smirnov one-sample test	Wilcoxon matched-pairs signed-ranks test	Mann–Whitney U test	Friedman two-way analysis of variance	Kruskal–Wallis one-way analysis of variance	1. Spearman rank correlation coefficient: r_s 2. Kendall rank correlation coefficient: τ
Interval	Normal curve and z	t test	t test	ANOVA	ANOVA	Pearson's product — moment correlation coefficient: r

It cannot be stressed too much that, even though the statistical analysis is necessarily performed at the end of a series of experiments, a consideration of statistical methods is a most important part of the *initial* design of an experiment. There is a liaison between experimental design and statistical analysis which, if properly used, will strengthen the validity of any conclusions that are drawn. It will also help to clarify the type of measurements that need to be made and even the number of experiments that need to be performed. By contrast, a poorly designed experiment cannot necessarily be 'salvaged' by an array of sophisticated statistical techniques; if the design has been inadequate the statistical analyses that can be performed validly and the conclusions that can be drawn might be more restricted than originally intended.

7.1 Deciding upon the most appropriate statistical test

This decision is based upon a number of considerations.

7.1.1 *Consideration one: the type of hypothesis being tested*
The general outcome of a series of experiments designed to test a hypothesis is a set of results of which one or more of the following questions is asked:

Question 1: Does the set of results (sample) fall within acceptable limits as far as some population is concerned — in other words, is the sample 'healthy' or 'normal'? (Note that this use of 'normal' relates to common, not statistical, parlance).

Question 2: Do two sets of results obtained following different treatments differ — in other words did the different treatments have any effect?

This question is asked by the experimenter when he compares control and experimental samples; it is the question most commonly asked.

Question 3: Do several samples which have received a variety of treatments differ from each other?

In some experiments more than one experimental procedure or treatment is involved; for example, more than one diet, drug, rate of electrical stimulation, means of assessment of mental alertness, etc. One possible means of analysis, if there are four treatments A – D, is to test all possible pairings of treatments (such as A *v.* B, A *v.* C, A *v.* D, B *v.* D, B *v.* D, C *v.* D. This is unsatisfactory because: (a) it is using each set of data more than once and; (b) as the number of tests increases, some pairings will be 'significantly different' merely due to the numbers of tests involved. This point is considered further in Section 7.2.1.

In these circumstances, the appropriate statistical tests are *analyses of variance*. These are so called because they divide the observed variation into that between groups and that within the groups. They then test if the variation between the groups is significantly greater than that within them. These analyses are able to deal with experiments where more than one factor is varied. Thus one might consider an experimental design in which the effects of two factors are considered. Consider the example where two changes are given simultaneously. One possible experimental arrangement is shown in Table 7.2.

Table 7.2 Format for two-factor analysis of variance

	Dose of drug								
Drug X	Low			Medium			High		
Drug Y	Low	Medium	High	Low	Medium	High	Low	Medium	High
Nine groups (*a*) to (*i*) each of *n* = 5	a_1	b_1	c_1	d_1	e_1	f_1	g_1	h_1	i_1
	a_2	b_2	c_2	d_2	e_2	f_2	g_2	h_2	i_2
	a_3	b_3	c_3	d_3	e_3	f_3	g_3	h_3	i_3
	a_4	b_4	c_4	d_4	e_4	f_4	g_4	h_4	i_4
	a_5	b_5	c_5	d_5	e_5	f_5	g_5	h_5	i_5

When the effect of more than one factor is being investigated the experimenter should endeavour to have equal numbers of results for each combination of dose levels of the two drugs, namely five in this example. (It is probably desirable that there should be more than a single value in each subgroup, to enable an estimate of within-subgroup variability to be made.) This method of analysis allows the effects of different dose levels of drug *X* and drug *Y* to be considered. In addition, any possible interaction between the drugs can be investigated. An example of interaction occurring would be if differences due to the dose of drug *X* existed only at some concentrations of drug *Y*. As before, a variety of tests exists (see Table 7.1).

Question 4: When the individuals comprising the sample are considered, is there any reliable relationship between a chosen pair of variables — in other words, do the two variables correlate?

Note that this question can also be asked of a pair of variables measured repeatedly in a single individual but that it is unacceptable to 'pool' results from a mixture of these two approaches (see Chapter 6). (That is if five individuals each produce four pairs of results, it is wrong to make n, the number of values in the sample, equal to 20. Either a separate assessment of correlation is made for each individual, $n = 4$ each time, or each individual donates one pair of data points — the average of the separate measurements — to produce a sample of size $n = 5$.)

7.1.1.1 *The statistical importance of standardizing the experimental conditions.* Such questions can be usefully posed only if the results have been obtained from viable experimental preparations under conditions that have been standardized as far as possible. As has already been described (in Chapter 3), the production of such conditions is the aim of the 'control' experiments and the results obtained from them are the control sample; these are then compared with the results obtained under experimental conditions (experimental sample) which must differ from the control experiments in one factor only. Note that with complex protocols, more than one type of 'control' is possible (see Figure 6.5). We stress that, in such cases, the experimental data must be obtained under conditions as identical as possible to those which enabled the control values to be measured. This can be difficult since the control or accepted value has often been measured some time earlier than the proposed set of experiments are undertaken. In the meanwhile, there might have been changes in the measurements or criteria that are used (for example, sensitivity of the assay, severity of illness before admission to hospital, social acceptability of possessing the factor or attribute, etc.). This problem can be a particularly difficult one in the case of retrospective studies (see Chapter 3). As with problems arising from inadequate laboratory measurement, statistics cannot rescue a poorly designed experiment which has inappropriate controls or experimental phases. The outcome of such experiments will be meaningless.

7.1.2 *Consideration two: level of measurement*
7.1.2.1 *What is the level of measurement?* The level of measurement can be described as nominal, ordinal or interval.

Nominal measurements divide results into classes or categories (for example, 'did die' or 'did not die' after drug treatment). All examples in any category are treated as being equal even if (in the example given) some might have died sooner than others after drug administration.

Ordinal measurements rank the results so that one distinguishes 'more than' from 'less than', for example, by reference to length of survival after giving a drug. However, there is no knowledge of how much one rank is greater than another, and adjacent ranks will be separated by different absolute amounts.

Interval measurements assess values against an agreed scale (for example, temperature, time, etc.). By such measurements it is possible to calculate *by how much* two measurements differ. Thus the difference between two values of, say, '10' and '12' is exactly one-fifth of that between '3' and '13'. For this reason, such measurements and the statistical tests that can be used upon them are called 'parametric'.

A further scale exists in which criteria applicable to an interval scale apply but, in addition, there is an absolute zero. For example, if one is measuring weight, then there is a true zero '0 g' and a weight of 100 kg is twice one of 50 kg. However, in the example of temperature, there is no true zero and thus a temperature of 20° C is not twice as hot as one of 10° C. A scale which possesses an absolute zero is called a geometric scale. The data measured on such a scale can be assessed by a parametric scale and also be described by a geometric mean.

7.1.2.2 *Level of measurement applied to a single sample.* This concept of the level of measurement can be illustrated by reference to the sample of Table 6.1. Some parameters of this sample are given in Table 7.3. We will consider the conditions that must apply for each description of this sample to be valid.

We start with the most sophisticated description, namely, the mean and standard deviation. These require measurements to be of interval status. For the mean to be useful, the individuals in the sample must be symmetrically distributed about it and, for the standard deviation

Table 7.3 Results derived from sample of Table 6.1

Level of measurement	Central tendency		Spread	
Interval	Mean	= 43.18	Standard deviation	= 20.43
Ordinal	Median	= 42	Interquartile range*	= 28−52
Nominal	Mode	= 52	Range	= 8−91

* excluding eight values at either extreme.

to be useful, the variability within the sample must be distributed in a normal (Gaussian) manner. This latter condition is sometimes not fulfilled, especially if ratios are being considered. A technique that can 'normalize' the data, that is, produce a distribution that is more Gaussian, is to take the logarithms of the values and to estimate the mean and standard deviation of these transformed data. The results can then be expressed in this format (Figure 7.1). At the end of any

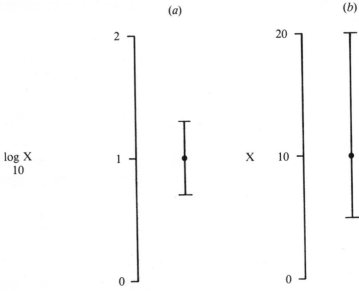

Fig. 7.1 Mean ± SD of sample; (*a*) expressed as 'logged' (normalized) results; (*b*) as expressed after transformation back to original formal ('anti-logging').

calculations, for illustrative purposes, the results are sometimes converted back to their original format (that is, 'anti-logged'). Note that this will produce a value for (mean + SD) which is further above the value for the mean than is the value for the (mean − SD) below it (Figure 7.1*b*). However, if statistical tests are to be performed upon such results, then the transformed data should be used.

Often, these strict criteria for calculating the mean and standard deviation cannot be met and so other levels of measurement are the only appropriate ones. The median and quartiles, for instance, are concerned with the relative size or *rank* of the individuals rather than their absolute difference. Thus, whereas rank 1 (the highest) must be better than rank 2, the amount of difference is neither stated nor required. Sometimes such measurements are all that can be achieved.

For example, if students are required to rank textbooks, the order of preference can be clear, but the highest rank might be better than the next highest rank by a trivial or an overwhelming amount. On occasion, subjective feelings are assessed by some scale. For example subjects might be asked to answer the question 'How tired do you feel?' by giving a number from 1 to 10 where '1' signifies 'not at all tired' and '10' signifies 'as tired as I have ever felt'. However, this approach does not convey an absolute meaning to a value of (say) '6'. Although it can be stated that '6' is certainly greater than '5', it is not possible to quantify the amount of difference. The calculation of a mean and standard deviation in these circumstances would be inappropriate: calculation of the median and interquartile (or a percentile) range would be more justified.

Sometimes, it is not even possible to rank individual results. Thus, given a list of characteristics required of students (for example, motivation, intellect, powers of concentration, honesty, dress-sense), one can do little more than agree or disagree with each requirement. To rank one category above another, for example, motivation > honesty, might be more an indication of the experimenter's prejudices than of anything else. In this case, the most popular choice — the mode — might be of most use.

7.1.2.3 *Level of measurement as applied to a number of samples.* These discussions of levels of measurement and the processes of 'upgrading' and 'downgrading' data apply not only to a description of a single sample but also when differences between samples and correlation between two variables are being investigated. It is the *lowest* level of measurement in a set of results that determines the appropriate statistical test to use (see Table 7.1). As will be seen later (in Section 7.3.2), tests involving interval data tend to be more powerful (that is, they require fewer results to show a significant change) than those involving categorization of results (using nominal data). Ranking tests (using ordinal data) are intermediate in power. This fact will also be important when consideration is given to the number of experiments that will need to be performed.

7.1.2.4 *'Upgrading' and 'downgrading' the level of measurement.* If a sample of data (see Table 6.1 for instance) has been measured as an interval scale and it shows a normal distribution — or the results have been 'normalized' by logarithmic transformation — then the experimenter can describe the sample in terms of its mean and standard deviation. However, there might be circumstances when he is more concerned with knowing the median, which divides the sample so that there are equal numbers of values above and below this parameter or he might wish to know the mode, the most commonly occurring value (see Table 7.3). The median is a useful description of the sample, for example, if the results are skewed and he does not wish to work with transformed data. The mode has value if he wishes to store only a limited stock of a commodity from the whole range that is available and wants this stock to be the most popular. Note that, to calculate the median, not all the information contained in the original measurements is being used; only the relative, not the absolute, size of each value is required. In the case of the mode, this restriction in the use of the available data is even more marked. This process of reducing the use that is made of the available data can be called 'downgrading'. By this process, interval data can be treated as ordinal or nominal data and results achieving an ordinal status can be treated as though measured nominally. To consider the data of Table 6.1, they might be considered as of ordinal status if they represented the scores of 33 individuals who were assessed for 'fitness' but on some arbitrary scale ranging from 0 (totally unfit) to 100 (totally fit). Such a scale could not have a strong claim to being equi-interval (see Section 7.1.2.1). If the results of Table 6.1 were presented in nominal format then a previous decision would have been taken either arbitrarily or (preferably) by reference to the scientific literature or to some theoretical model. For example, values might be categorized as:

    86 +   very good
    66 − 85 good
    36 − 65 moderate
    16 − 35 poor
     0 − 15 very poor.

(Note that the categories do not have to be of equal size, though there is an argument for dividing them so that the population as a whole will divide approximately equally between the categories.)

The results of Table 6.1 would then be recorded as:

    very good   1
    good   4
    moderate   14
    poor   12
    very poor   2

It is important to realize that the processes of 'upgrading' results — those of converting nominal data to ordinal or interval status or of converting ordinal data to interval status — are unjustified. They will almost always entail making unwarranted assumptions about the data. Thus to convert data of ordinal to interval status one has to make the assumption that the ordinal results are measured on an equi-interval scale. The process of 'upgrading' nominal data is even more unjustified since there is no way of knowing where in a particular category an individual result was. Thus the 14 'moderate' results (above) could all be as high as 65, as low as 36, equally spread between 36 and 65, etc.

7.1.3 *Consideration three: paired or unpaired experimental design*
The basic difference between these two designs has been discussed in Chapter 3. Briefly, in unpaired designs, individuals are randomly assigned to the different groups or samples; individual differences are 'averaged out'. With paired designs, individual members of matched pairs are randomly assigned one to each group. Differences between each pair are then measured. An alternative and better form of matching, possible in some studies, is when the same individual is studied twice, under control and experimental circumstances. This is better than matched pairing because, in the latter case, one can never match individuals *exactly* or be sure that one is matching for all the appropriate factors. However, when experimental manipulations are irreversible, or if the effect of a drug takes too long to wear off, then the same individual cannot be used. When more than two groups are being considered, as during studies involving analysis of variance, then matched sets of animals are required (for example, sets of four if there are four groups A – D).

The statistical implication of paired and unpaired designs can be understood by reference to the data of Table 7.4. The values are to be considered as achieving interval status.

Consider, first, the experimental design by which groups A and B are unpaired. The mean and standard error of the mean of each sample are shown at the foot of the table. Our intuition that the variability within each sample is large enough to render non-significant any difference between the means is, in fact, borne out by use of Student's unpaired *t* test.

However, let us suppose instead that the results were obtained from paired or matched samples; that is, each row in Table 7.4 refers to results from the same animal or from its matched pair under two circumstances,

Table 7.4 Hypothetical data (for details see text)

	Sample		Difference
	A	B	B − A
	8.06	10.00	1.94
	8.40	9.14	0.74
	9.67	10.43	0.76
	9.91	8.96	−0.95
	10.31	11.26	0.95
	10.45	12.60	2.15
	10.62	10.61	−0.01
	10.78	11.81	1.03
	11.58	11.95	0.37
	11.92	13.09	1.17
	12.20	12.52	0.32
	12.23	12.48	0.25
	12.46	13.71	1.25
	12.99	12.13	−0.86
	13.23	12.07	−1.16
	13.54	13.73	0.19
	14.09	14.66	0.57
	14.17	13.73	−0.44
	14.44	15.89	1.45
	15.10	14.97	−0.13
	15.50	17.24	1.74
	17.05	16.79	−0.26
Mean	12.21	12.72	+0.50
SE	±0.50	±0.48	±0.20
n	22	22	22

A and B. We can consider the difference between the two results as is done in column B − A. The mean difference is now considerably greater than its standard error and this suggests that the difference might be significant; use of Student's paired t test confirms this supposition.

Reference to Table 7.1 will show that statistical tests for use with paired data exist at all levels of measurement and for two or more samples. In all cases, the experimental design using paired data is more powerful since it reduces the variation due to interindividual differences.

7.2 Interpreting statistical tests

As a result of the previous considerations the experimenter will have been able to choose the test that is most appropriate for his requirements and data. When a statistical test is carried out it requires calculation of a particular statistic (for example t in Student's test, U in the Mann–Whitney test or r for the correlation coefficient). An appropriate table is then consulted which gives an estimate of the likelihood of obtaining such a value for the statistic — or one more extreme — assuming that the null hypothesis were applicable. This likelihood is expressed as a P-value ranging between 0 and 1.

7.2.1 *The null hypothesis and its rejection: type I and type II errors*
The null hypothesis is that of no difference. For example it would indicate: that the sample has been drawn from the main population; that the two (or more) samples have been drawn from the same population, that is, the treatment has had no effect; and that there is no relationship between the two variables. When P is small enough, the null hypothesis is rejected, and the difference is then said to be 'significant'. The value of P when this rejection takes place is arbitrarily decided (but should be done so before the statistic is calculated), but is generally $P < 0.05$ or $P < 0.01$. These two values give rise to the expressions rejecting the null hypothesis at the 5 per cent level or at the 1 per cent level, respectively. Clearly, the '1-in-20' or '1-in-100' chance can occur, so there is the possibility of rejecting a null hypothesis falsely. This is termed a type I error. The chance of this happening will depend upon the significance level that has been used in deciding whether or not to reject the null hypothesis. Thus, it will be greater if a value of $P < 0.05$ rather than $P < 0.01$ is chosen.

The chance of making a type I error increases rapidly when multiple comparisons are involved. Consider five groups, A – E, and attempts to distinguish between them. If all possible pairs are considered (A *v.* B, A *v.* C, A *v.* D, A *v.* E, B *v.* C, B *v.* D, B *v.* E, C *v.* D, C *v.* E, D *v.* E) then ten comparisons are made. If the chance of making a type I error in any individual comparison is, say, $P = 0.05$, then the chance that no such errors are made in ten comparisons is 0.60. That is, there is a 40 per cent chance of making at least one type I error in this analysis. It is for this reason that, in such circumstances, multiple comparisons are less desirable than analysis of variance techniques (see Section 7.1.1).

Another problem is that, if the change is a small one, it might be 'missed', that is a (real) difference is falsely regarded as being a non-significant one and the null hypothesis is not rejected. It is important to realize that not being able to show that a difference exists between two groups, etc. is *not the same* as stating that there is no difference between them. This is a type II error and the likelihood of this occurring increases as the value for P required to reject the null hypothesis falls. In other words, if the number of experiments is constant, the chance of making a type I error can be decreased only by increasing the chance of making a type II error and *vice versa*. However, both errors are less likely to be made if the number of experiments is increased (see Section 7.3.1).

7.2.2 *One-tailed and two-tailed tests*

Often, the investigator is concerned only with whether or not to reject the null hypothesis; the direction of change is of no interest to him and extreme results in either direction will enable the null hypothesis to be rejected. In these circumstances, it is the P value from a two-tailed statistical test that is required when the decision to reject or not to reject is taken. If the null hypothesis is rejected, then the alternative hypothesis—that there is a difference between the sample and the population, etc. — can be accepted. Sometimes, the alternative hypothesis might prescribe a direction of change and then a one-tailed statistical test can be used. For example, the ability of a drug to *increase* sleep might be investigated. Note that, if the drug produced large *decreases* in sleep, the alternative hypothesis would not be supported. However, if the alternative hypothesis were merely that a change exists (and the direction was not prescribed), then the null hypothesis could be rejected and the alternative hypothesis could be accepted on this evidence. Often, it is difficult to be *sure* that a one-tailed test is appropriate. Using an 'antihypertensive drug' and proposing an alternative hypothesis that the drug produces a *fall* in blood pressure (and hence using a one-tailed test) might seem acceptable. However, this would not be the case if the drug were still being tested and its ability to lower blood pressure had not yet been established beyond doubt; an alternative hypothesis postulating a difference without regard to direction (and hence the use of a two-tailed test) would be more appropriate.

7.3 **How many experiments need to be performed?**

Obviously the experimenter does not wish to waste material, effort, time and money on performing an unnecessary number of experiments; equally it is most frustrating to obtain results that do not achieve statistical significance because of groups of insufficient size. The exact number of experiments that needs to be performed cannot be predicted exactly but some guidelines do exist.

7.3.1 *Decreasing type I and II errors*
The larger is the experimental group, the more representative it will be, that is, the less likely is it to contain 'freak' results. An implication is that a false rejection of the null hypothesis (type I error) is less likely to occur. This position will hold whatever the level of measurement or type of question being asked and whether the data are paired or unpaired.

Increasing sample size increases also the likelihood of demonstrating the significance of this difference or correlation; again, this applies at all levels of measurement. In other words, the likelihood of falsely accepting the null hypothesis (making a type II error) decreases with increasing the amount of data involved.

7.3.1.1 *Nominal data.* Consider, for example, the contingency table
derived from hypothetical nominal data (Table 7.5a). This gives a value of $\chi^2 = 2.8$ and, from tables, $P = 0.09$. This difference is not significant, therefore. (Yates's correction has been used.)

If the *same* distribution is retained but with twice the number of results (Table 7.5b), then $\chi^2 = 6.7$ and $P = 0.009$. That is, the difference is now a 'significant' one. (Again, with Yates's correction.)

Table 7.5 Hypothetical data (for more details, see text)

		Factor X	
		Present	*Absent*
A.			
	Group A	6	12
	Group B	12	6
B.			
	Group A	12	24
	Group B	24	12

7.3.1.2 *Ordinal data.* Consider the example in which two groups, A and B, are ranked and assessed by the Mann–Whitney U test. Ranking them as:

$$A,A,A,B,A,B,B,B \quad (n_A = n_B = 4)$$

gives $U = 1$ and a two-tailed P value of 0.058. If the same distribution is doubled in size, we can have the ranking:

$$A,A,A,B,A,A,A,B,A,B,B,B,A,B,B,B \quad (n_A = n_B = 8).$$

This gives $U = 10$ and a two-tailed P value of 0.02.

Ranking the larger sample differently, we could have:

$$A,A,A,A,A,A,B,B,A,A,B,B,B,B,B,B \quad (n_A = n_B = 8).$$

Here, $U = 4$ and the two-tailed P value is 0.002.

Again increasing the sample size, but not changing the distribution of the results, has led to a significant difference no longer being missed.

7.3.1.3 *Interval data.* With Student's unpaired t test, as sample size increases, the pooled sample variance:

$$s^2 \left(\frac{1}{n_1} + \frac{1}{n_2} \right)$$

falls due to increases in n_1 and n_2. As a result, the value of t in the equation:

$$t = \frac{| \bar{x}_1 - \bar{x}_2 |}{\sqrt{s^2 \left(\frac{1}{n_1} + \frac{1}{n_2} \right)}}$$

will increase. Further (see tables), the value for t required to reject the null hypothesis falls with increasing degrees of freedom, where these equal $(n_1 + n_2 - 2)$. These two factors both contribute to the decreasing chance of making a type II error with increasing sample size.

7.3.1.4 *Correlation.* The concept that, as the sample size increases so the value of a statistic that must be reached if the null hypothesis is to be rejected tends to decrease, applies also to correlation. Thus, for any correlation coefficient (provided its value is not zero), the greater is the number of values that contribute to it, the more likely is such a value to be significantly different from zero. This can readily be ascertained from the relevant tables. As an example, a Spearman's rank correlation coefficient of 0.7 is not significant ($P > 0.05$) if the number of pairs of values is 6, but with 12 pairs of values $P = 0.02$ (two-tailed test).

7.3.2 *The power of different statistical tests*

As stated already (in Section 7.1.2.3), statistical tests using data of interval status are most powerful, that is, they require the smallest sample size to show a significant difference. The tests using nominal data are least powerful in this respect and ordinal tests are of intermediate power. This result is perhaps not surprising when one considers the extra information contained in interval measurements and the stricter criteria that have to be applied before interval tests are suitable.

These general ideas can be understood by reconsidering the paired data of Table 7.4. Assuming them to be of interval status, it was mentioned (in Section 7.1.3) that the differences between samples A and B were significant. In fact, by Student's *t* test the *P* value (two-tailed) is less than 0.02. If, by contrast, the data are treated as of ordinal status, then Wilcoxon's matched-pairs signed-ranks test (see Table 7.1) gives a two-tailed *P* value of $0.02 < P < 0.05$. You will notice that this would appear to indicate a less significant difference between groups A and B. Finally consider the possibility that the data had been recorded in categories instead, values less than, or equal to, 12.5 being classified as 'high', the rest as 'low'. The results would then have been recorded as in Table 7.6 and can then be used to calculate χ^2 in the manner suitable for the McNemar test (see Table 7.1). The value of χ^2 is too low to enable the null hypothesis to be rejected and a type II error would have been made. (Strictly, the number of values in Table 7.6 is rather small, so the binomial test is more appropriate; with this test also, the null hypothesis cannot be rejected.)

Thus, in this case, the null hypothesis cannot be rejected whereas in the cases when ordinal or interval measurements were used, then it

Table 7.6 Data derived from those of Table 7.4

		After treatment (group B)	
		High	Low
Before treatment	high	7	2
(group A)	low	4	9

could be at the 5 per cent level or less. This example shows the advantage to be gained by making measurements that reach ordinal or even interval status, if this is possible. Also, by implication, the ordinal and nominal tests require progressively more results to avoid making a type II error. Both of these factors are important when the experiment is being designed. This example also shows the pitfalls that might arise if measurements that have achieved interval status are 'downgraded' into ordinal or nominal status.

7.3.3 *The minimum number of experiments*

There is a minimum number of experiments that must be performed for a significant difference to be demonstrated; this minimum is larger for non-parametric tests. For example, consider the minimum sample size required to reject the null hypothesis using the binomial distribution. Assume that the chance of possessing a characteristic is 0.50 (and so the chance of not possessing it is 0.50 also) then the most extreme result that is possible is either that all members of the sample possess the characteristic or that none does.

Table 7.7 shows the probability of obtaining a sample, all members of which have the same characteristic. Unless the sample size is greater than four, the chance of this happening is sufficiently high for it not to be regarded as significant at the 5 per cent level. This is with a one-tailed test in which the direction of extremeness ('all possess' or 'none possesses') is stated beforehand; with a two-tailed test, when *either* extreme result is acceptable, a minimum sample size of six is necessary.

Notice also that with a single 'aberrant' value (that is, one with the opposite characteristic to the others), the minimum sample size that

Table 7.7 Effect of sample size and aberrant values upon *P*-values (for more details see text)

Sample size	Number of 'aberrant' values	One-tailed test	Two-tailed test
6	0	0.02	0.03
5	0	0.03	0.06
4	0	0.06	0.12
9	1	0.02	0.04
8	1	0.04	0.07
7	1	0.06	0.12
6	1	0.11	0.22

is required to show a significant difference increases to eight (one-tailed test) or nine (two-tailed test).

With an interval test, by contrast, a significant difference from a given population distribution can be found with smaller samples. In the limiting case a *single* value can be established as significantly different from the population at the 5 per cent level if it differs from the population mean by more than 1.96 × SD. (This assumes that we know the population SD and that the distribution is normal.)

This greater power of parametric tests applies also to tests of correlation and to a comparison between two samples. For example, with nominal data and by the use of Fisher's exact test, it can be shown that, to reject the null hypothesis, minimum sample sizes of four are required in each group to achieve a P value less than 0.05 (two-tailed test). This is with the most extreme result (for example, all four of sample A possess the factor, but none of the four in sample B); with even a single 'aberrant' value, the sample size must be greater. Similarly, with an ordinal test (Mann−Whitney U test), minimum sample sizes of four in each group (A and B) are required to achieve significance at the 5 per cent level (two-tailed test), again with the most extreme result:

$$A,A,A,A,B,B,B,B$$

By contrast, a significant difference between two sets of interval data (assuming the other criteria are met, of course) can be obtained with each sample consisting of only two values ($n_1 = n_2 = 2$) if the difference between the sample means ($\bar{x}_1, \bar{x}_2$) is sufficiently large in comparison with the pooled sample variance (s^2), that is, if

$$\frac{|\bar{x}_1 - \bar{x}_2|}{\sqrt{s^2 \left(\frac{1}{n_1} + \frac{1}{n_2}\right)}} \geqslant 4 \cdot 30 \qquad \text{(two-tailed test, d.f. = 2)}$$

7.3.4 Summary

In summary, therefore, increasing the number of experiments increases the likelihood of establishing a significant difference or correlation where one exists (that is, of not making a type II error). The exact number depends upon the power of the test and this tends to be lower for nominal data (that is, more values are required) than for interval data. However, the advantage of ordinal and nominal tests is their wider applicability, being less stringent than interval tests in the criteria that the data must satisfy. There is also the obvious advantage in increasing the number of experiments in that the effect of extreme or freak values will be lessened overall and with it the chance of drawing a false conclusion from the results. Clearly, the number of values required to achieve these aims cannot be exactly specified, but there are minimum sample sizes below which, on theoretical grounds, significant differences can never be demonstrated. It is important to realize that many of these issues are to be resolved whilst the experimental protocol and types of measurements to be made are being planned; statistics has a place in the *design* of an experiment as much as in an assessment of its results.

Chapter 8

Drawing conclusions

The aim of experiments has been to test a prediction made by a hypothesis. Before the outcome of the test can be assessed, the following assumptions must be made:

(1) The preparation was physiologically healthy throughout the experiment.
(2) The protocol was adequate both with respect to the experimental regimen and the choice of control phases and experiments.
(3) The apparatus that was used to perform any experimental manipulations or record the results was appropriate for the task and calibrated and used correctly.
(4) The calculations that were carried out upon the results were appropriate to the test under consideration and took into proper account the control experiments.
(5) The correct statistical analyses have been performed.

8.1 Assessment of the experimental results

Clearly, the results are the key part of the experiment. When obtaining them, it is wise for the experimenter to know of some of the problems that he faces.

8.1.1 *Is the recorded effect due to the stimulus, or would it have occurred independently?*

The experimenter will have recorded the effect of a particular intervention, the stimulus, but how is he to show that the response was due to the stimulus and would not have occurred without it?

For example, does his result merely show a deterioration of the preparation, especially in a long experiment in which extensive surgery has been carried out. In an *in vitro* preparation, it may be because of inadequate perfusion or an inappropriate solution bathing the preparation. There may be other effects due to the passage of time but which are not due to a deterioration in the preparation. For example, in long-term studies on neonates, there will be changes in their level of development and, at the other end of life's scale, there may be effects of ageing. There are also going to be changes over a rather shorter time scale; for example, following nerve section, there may be some nerve regeneration or, following the implantation of a recording device such as is used for recording blood flow or interstitial fluid pressure, growth of tissue may alter the calibration of the device. Fatigue may also be a problem in exercise studies or in studies on isolated muscle. The changes observed may be in response to some factor other than the stimulus, for example, changes in the environmental temperature or the amount of oxygen available to the tissue. The change might reflect the fact that the responses to drugs may also alter with repeated injection of the drug, the tissue becoming less sensitive to it.

In a well-planned experimental protocol, the experimenter ought to be able to answer these points. He will be able to do this by reference to his control phases and his control groups of animals and the sequence of control and experimental phases. That is, if the experimenter has carefully designed his experimental protocol (see Chapter 3), he should be able to decide if the recorded response does result specifically from the stimulus rather than from some other factor that would have occurred spontaneously.

8.1.2 *Is the response a specific or a non-specific result of the stimulus?*

In many experiments, the experimenter is interested in the effect of a stimulus on one particular tissue or group of receptors or cells. How then does he show that the effects he has recorded are due to a specific effect and are not merely part of a more widespread change? Many drugs, hormones and neurotransmitters have very widespread and rather non-specific effects, and it may be that the recorded effect is due to one of these rather than due to the specific effect aimed for by the experimenter. Non-specific stimuli may also be a problem in neurophysiological studies when attempting to stimulate a particular group of nerve cells (without influencing the activity of nerve fibres passing through the site of stimulation), or a single type of receptor. Further, particularly with a very intense stimulus, there is the problem that a non-specific, 'alerting' or 'stress' response might be evoked.

Many of these problems of stimuli producing general effects due to the non-specific nature of the stimulus can be surmounted by the use of discrete stimuli as discussed in Chapter 5. However, one must then try and eliminate the possibility that the stimulus is having widespread effects. It may be possible to design the experimental protocol so that the stimulus is delivered in a number of ways. (Thus one might use a variety of techniques to stimulate a receptor or to change blood chemistry.) If the measured response was non-specific, then it would be expected to be seen regardless of the way in which the stimulus was given. By contrast, a specific response to a stimulus would be expected to appear only in response to one particular circumstance.

8.2 Comparing the experimental result with the prediction made by the hypothesis

In general, the outcome of the experiments will either agree or disagree with the prediction made by the hypothesis. Each outcome will be considered in turn.

8.2.1 *First outcome — the result agrees with the prediction*
Certainly, this is the result if the hypothesis is correct, but, unfortunately, such a result can be obtained for other reasons.

Consider, for example, the apocryphal story of the foolish man who wished to show that an insect heard through its legs. To test this, he taught the insect to dance on his verbal command. He then removed the legs of the insect; after removal of its last leg, the insect did not dance on his command. He concluded that the experimental result supported his hypothesis.

A similar kind of interpretive problem often arises when experimental results are considered (though in a rather more subtle form, of course!). It arises if the experimental interventions that are used or measurements that are taken are not *direct and immediate* tests of the hypothesis.

Thus, where the hypothesis has stood up to its first test, the next stage is to devise a more stringent test. Two related approaches exist. With the first, the aim is to remain within the framework of the original hypothesis but to limit some of the possible alternative explanations by further experiment; with the second approach, the test is made more stringent by becoming a more 'direct' test of the hypothesis.

In the following examples, the initial results agreed with predictions made by a hypothesis, but later, more detailed tests of the hypothesis produced contradictory results. Consider first the hypothesis that memory formation requires RNA synthesis. This is supported by the observations that injection of RNA improved learning and that injections of RNAase, a hydrolytic enzyme that breaks down RNA, impaired memory formation.

However, the changes to RNA levels might have influenced many other aspects of neuronal activity. In order to limit these other possible explanations, it becomes important then to establish that memory is associated with specific, not general, changes in RNA. In this example, this requires the demonstration of the transfer of specific, rather than general, memories via RNA molecules which have a chemical structure that is specific for the memory being transferred. (This illustrates the first approach.)

Let us now consider the hypothesis that the loop of Henle concentrates urine. Results supporting this are that animals with loops of Henle could produce urine more concentrated than plasma, and that the longer the loops the higher was the maximum concentration that could be achieved.

However, when a more stringent test of the hypothesis was applied, it was found that the sampled fluid entering the distal convoluted tubule was hypotonic to plasma at a time when concentrated urine was being produced; thus, the loop of Henle, though required for urine to be concentrated, did not itself concentrate the fluid. (This illustrates the second approach quite clearly.)

To take a further example, let us consider the hypothesis that males are less intelligent than females. Results supporting this are that more males than females fail university examinations, and the males' average mark is lower than that of females.

However, the males might have greater distractions than the females (you decide upon possible factors!). This result implies a fault in the original statistical design, since the groups must be paired also for 'motivation', 'distractions', etc. (see Section 3.2). However, you cannot match for a factor you do not initially know to be important.

This last example illustrates a problem that recurs continually when experiments are performed. Unlike the first two cases when the test of the hypothesis was modified in some way (in example 1 by eliminating alternative explanations, or in example 2 by testing the hypothesis more directly), in this case, the protocol has to be changed so that the same test can be carried out more correctly. In other words, the advance in knowledge produced by the first experiments enabled the protocol to be redesigned to eliminate statistical and methodological errors which, unknown to the experimenters, existed initially. Clearly, this is an important aspect of testing a hypothesis under stringent conditions.

8.2.1.1 *Interpreting a correlation between two variables.* There are some occasions when correlations are high but the result is inevitable and can be predicted from theory. Consider the observation that there is a correlation between glomerular filtration rate (GFR) and sodium reabsorption by the kidney (R_{Na}). This correlation is inevitable since the two variables being plotted against each other share a common factor. Thus:

$$\text{sodium reabsorption} = \text{sodium filtered} - \text{sodium excreted}$$

Now $Na_{FILT} = GFR \times P_{Na}$ where P_{Na} = plasma concentration of sodium

Na_{FILT} = filtered load of sodium

and $Na_{exc} = U_{Na} \times \dot{V}$ where Na_{exc} = sodium lost in urine

U_{Na} = urine concentration of sodium

$\dot{V}$ = urine flow rate

Therefore $R_{Na} = GFR \times P_{Na} - U_{Na} \times \dot{V}$

This could be investigated by measuring R_{Na} and GFR in a number of subjects and plotting these two variables against each other. However, the experiments are unnecessary since one is comparing (GFR $\times$ $P_{Na} - U_{Na} \times \dot{V}$) with GFR. The term GFR is common and so some degree of correlation is inevitable. Further, since urinary loss of sodium is much smaller than filtered load, and plasma sodium varies only slightly, the correlation is likely to be very high as the investigation approximates to a comparison of

(GFR $\times$ constant factor $-$ negligible factor) with GFR.

Such a correlation can be deduced from theoretical considerations. It involves valuable concepts, but the experiments do not require to be performed. Sometimes, such inevitable correlations arise because of 'indices' or other 'factors', the use of which obscures their components. For instance, consider an investigation of the relationship between the partial pressure of nitrogen in alveoli and ventilation rate.

Suppose that the investigator believed (quite reasonably) that the response to nitrogen depended on the 'sensitivity' of the respiratory system.

The relationship being investigated is then between

$$\dot{V} \text{ and } P_{N_2} \times S$$

where

> $\dot{V}$ is the ventilatory rate
> P_{N_2} the partial pressure of nitrogen in the alveoli

and

> S the 'sensitivity' of the respiratory system.

This appears to be a thoroughly acceptable hypothesis and mathematical test of it. However, suppose, further, that the experimenter (again, apparently reasonably) had defined 'sensitivity' (S) as the gradient of the line relating ventilation ($\dot{V}$) to partial pressure of CO_2 (P_{CO_2}), that is:

$$S = \frac{\dot{V}}{P_{CO_2}}$$

In fact, combination of the two equations indicates that this investigation is then between $\dot{V}$ and $\dfrac{\dot{V}}{P_{CO_2}} P_{N_2}$. Again, since $\dot{V}$ is common to both variables, some correlation between them is inevitable. The moral in this case is that, where possible, the derivations of 'constants', 'indices', etc. should be included in any equations in which they are used.

8.2.1.2 *The link between correlation and causality.*

Often a hypothesis predicts that, because two variables are linked in some way, changes in one variable will be associated with changes in the other. This is true, but the opposite line of reasoning is not necessarily so. Thus, it is a fallacy to believe that, because two variables are correlated, they are therefore linked in some causal way.

Consider, for example, the observation that during the daytime plasma insulin concentration and urinary flow rates are both raised. If values during the course of a 24 h period were compared, there would exist a positive correlation between these two variables. However, it would be wrong to infer that changes in either were *responsible* for changes in the other; rather, both result from another cause, namely the effects of meals.

To take another example, it has been observed that people who snore have a higher incidence of heart disease. Thus again the two variables show a positive correlation. However, it may not be that snoring causes heart disease. In fact, a more likely explanation is that the group who snore tend to be fatter or smoke more heavily or have more respiratory problems, all of which would predispose them to heart disease. Again, it may be possible to differentiate between these two alternatives by suitable pairing (see Chapter 3) of the groups for obesity, smoking habits and the incidence of respiratory disorders.

At a more formal level, a causal link between two variables, P and Q, is shown in Figure 8.1, model A, whereas the other possibility — both are influenced by a common factor, X — is shown as model B. In both cases, one would predict that P and Q would be correlated. The problem, therefore, is in deducing information about causality from the results of correlation studies. One can never *prove* in a philosophical sense that a change in P causes a change in Q. In practice, the hypothesis that P directly influences Q will stand — if there is this kind of experimental evidence or sound theoretical reasons in its favour — until it can be disproved.

If model A is correct, then P and Q *must* correlate; but the position becomes more complex when it is realized that if P takes T minutes to produce its effect on Q, then the correlation to be expected is between P at any time (t) and Q at time $t + T$. When T is unknown, then a

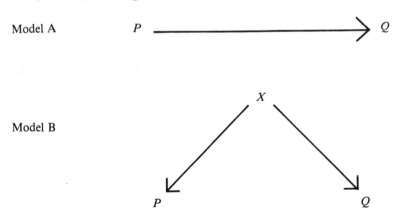

Fig. 8.1   Two possible models to explain a correlation between P and Q. In model A, P directly affects Q whereas in model B, both P and Q are affected by another variable (X).

series of correlations between P and Q can be calculated with different time lags. That is, a series of values of P is compared with the values of Q obtained 1, 2, 3, etc. units of time later. The unit of time could be from milliseconds (neural activity) to days (hormones) or even years (growth). The only sure disproof of the model is if P correlates with *earlier* values of Q. These points have been considered already in connection with the time-courses of cutaneous stimulation, sweating and deep body temperature in Figure 6.7. Model A is also amenable to further testing since changes (or prevention of normal changes) to P have predictable effects upon Q, but changes to Q have no influence upon P.

With model B, the correlation between P and Q might be expected to be less than with model A, since it arises indirectly. The time-lag between P and Q which gave the highest correlation would depend upon the time-courses of the processes by which X affects P and Q; it would be equally likely that the lag producing the highest correlation coefficient would be negative (that is, values of P would correlate with *earlier* values of Q) or positive (values of P would correlate with *later* values of Q). Changes in neither P nor Q should affect each other in the case of model B.

Clearly, whichever model applies, for the experiments to be of physiological relevance, changes in P (and X in model B) should always be within the 'physiological range', a point that will be discussed later.

8.2.2 *Second outcome — the result does not agree with the prediction*

If the hypothesis is wrong, then this is the result that would occur, but the result can occur even if the hypothesis is correct. There are two reasons for this. The first is redundancy. It is common knowledge that any physiological variable is controlled by a number of homeostatic mechanisms and that physiological processes act via (or are affected by) multiple pathways. The advantage of this redundancy from the viewpoint of the animal or plant is obvious, as destruction of one mechanism can be compensated for by the presence of other homeostatic mechanisms, but from the viewpoint of an experimenter, it poses a problem. The problem is that if part of a system which possesses a high degree of redundancy is removed by ablation, sectioning, etc. then the remainder will take over the role of the missing part. (This might account for the plateau values in Fig. 6.9 being equal both before and after nerve section.) As further examples, if a single carotid body is rendered functionless, three other peripheral chemoreceptor areas and the central chemoreceptors are still present to monitor and so control ventilation and gas tensions in the blood and cerebrospinal fluid. Similarly, the source of 'ergoreceptors' — neural inputs from an exercising limb — has often been investigated by cutting nerves from the muscle, tendon, ligaments, cartilage or surrounding skin, etc. If, as is intuitively likely, the CNS makes use of all ascending information, removal of one small part is unlikely to make much difference except possibly transiently.

There is a second and related reason why findings that disagree with predictions do not necessarily disprove a hypothesis. When attempts are made to produce a change in some physiological system, homeostatic mechanisms will 'correct' any imposed 'anomalies' by negative feedback mechanisms. Thus, if as a result of stimulating or cutting an afferent neural pathway from an exercising limb the respiratory responses to exercise *were* modified, then chemoreceptors would be likely to correct any change in ventilation fairly soon. That is, any effect due to stimulation or removal is likely to be transient rather than of a steady-state nature. (Note that this was illustrated in Figure 6.9.) In other words, the lack of response — certainly in anything other than the short term — might be misinterpreted to mean that the pathway under consideration was unimportant. As a further example, the effect upon heart rate and vascular smooth muscle of noradrenaline is far less marked and shorter-lived when injected intravenously rather than added to an isolated heart or artery preparation. It would be incorrect to infer from this that noradrenaline acted through different mechanisms *in vivo* from *in vitro*; in the former case, metabolic removal is more rapid and homeostatic mechanisms are present to oppose any changes resulting from direct effects upon the heart and blood vessels. (However, this *is* a reminder that *in vitro* results can differ from those observed *in vivo*. It also illustrates the concept (see Chapter 2) that the results of *in vitro* experiments are influenced less than those of *in vivo* experiments by changes brought about by homeostatic mechanisms.)

However, it would be wrong to infer that no such correcting mechanisms existed in *in vitro* preparations, rather they exist at a more local level. Thus, if an isolated organ is being perfused by a medium containing a vasoconstrictor drug, there seems to be no reason to believe that, should tissue ischaemia begin to arise, local changes in arteriolar tone will not occur reflexly to combat the abnormal partial pressures of CO_2 and O_2. Further, if isolated cells or tissue slices are bathed in a medium to which some substance is added, it is likely that the preparation will respond by metabolizing or sequestrating the added substance. That is, the size of the stimulus will change.

In summary, therefore, the effect of multiplicity of homeostatic mechanisms is that many stimuli appear to be ineffective or only transiently effective.

Two major approaches have been used in attempts to overcome these difficulties. The first is to control all pathways or mechanisms except the one under consideration so that the others continue to pass on 'no change' or 'normal' information. At the moment, this has often proved to be technically too complex to perform satisfactorily. However, one area where the problems have been overcome is in assessing quantitatively the relative contributions of CO_2, O_2 and acid as stimuli for the chemoreceptor reflexes shown by the whole animal. To assess the role played by any one of CO_2, O_2 or acid, it alone must be changed whilst the others must be maintained constant in spite of reflex changes in ventilation. Later, the relative importance of the peripheral and central chemoreceptors in producing these whole-body responses was assessed. For these experiments, the central chemoreceptors were perfused with an artificial cerebrospinal fluid (c.s.f.) so that the composition of blood and c.s.f. with respect to O_2, CO_2 and acid could be controlled independently. More recently, the reflex effects produced by the aortic and carotid bodies have been distinguished by perfusing one, but not both, set of peripheral chemoreceptors.

The second way in which attempts have been made to overcome the problems that arise due to the compensatory activities of homeostatic mechanisms is to deliver a stimulus that is so large that it 'overloads' the homeostatic system. By observing the abnormalities that result, one can infer the means by which homeostasis is normally achieved. As examples of this approach, one could cite experiments to investigate the effects of a protein-free or iodine-free diet, of severe heat or altitude or of maximal stimulation to a particular nerve. Related approaches are to consider the changes that occur under 'unusual' circumstances such as diving, living in outer space, pregnancy, etc.

However, it might be argued that these conditions share the property that they are all 'stressful'. As a result, it might be considered that in these experiments, a response to 'stress' rather than to an exaggerated homeostatic response is measured. As a result, therefore, and in spite of the importance of understanding 'trench-foot', 'mountain sickness', 'jetlag syndrome', 'grey-out' and 'the bends' from the viewpoint of occupational medicine, the usefulness of such techniques in understanding the *normal* physiology — of a standard man at rest at sea-level — is not necessarily clear. Put differently, an 'unphysiological' —that is, abnormal—circumstance might produce an 'unphysiological' response, rather than an exaggerated physiological one. For example, one would not readily predict the periodic 'flushing' of the hand that occurs when it is placed in iced water from a knowledge of its response to immersion in cool water; nor can the metabolic changes in an animal during the 'ebb phase' of its response to trauma — in which insulin insensitivity and inability to metabolize glucose exist — be easily predicted on the basis of either the subsequent 'flow' phase or preceding 'fight and flight syndrome'.

8.2.2.1 *Learning from 'wrong' results*

Even if an observation does not support the hypothesis to be tested, it may in itself provide a clue either to a different solution to that problem or to an entirely different problem. Many important scientific discoveries have been made by scientists considering observations (either made in their own laboratories or reported in the literature) in a different context—lateral thinking. Two well-known examples are the realizations of the potential medical usefulness of penicillin (when the mould producing this unintentionally gained entry into some bacterial cultures), and of the immunological implications of the observation that dairymaids did not suffer from smallpox.

Further examples are:

(1) Continued injections of aldosterone, although initially stimulating sodium reabsorption in the kidney, ultimately led to a decrease in this stimulation (the 'sodium escape' phenomenon). This might be interpreted to indicate the presence of a sodium-losing (natriuretic) factor.

(2) Patients have been found whose blood clotted too slowly in contact with a foreign surface, but in whom clotting was normal in the presence of 'a potent brain extract'. This led to the discovery of the 'intrinsic pathway' of coagulation.

(3) The effects upon glucose metabolism of alloxan-induced diabetes and pancreatectomy were not identical. This is now known to be due at least in part to pancreatic α-cells and their secretion of glucagon.

8.2.3 *Third outcome — results open to different interpretations*

An unlucky experimenter, using a physiologically normal preparation, can produce a stimulus which is both adequate and specific, can take recordings using suitable equipment which are free from, for example, movement artefacts and without any bias in selection and yet still his results may be open to conflicting interpretations.

Let us take, for example, the effect of stimulation of the carotid body on the heart rate of an anaesthetized animal. Either an increase or a decrease in the heart rate can be observed. The differences will depend upon whether or not ventilation was allowed to alter. Stimulation of carotid chemoreceptors by stimulating ventilation will increase the heart rate. However, if the ventilation is controlled then stimulation of carotid chemoreceptors will result in a *decrease* in the heart rate. It will depend upon the purpose of the experiment whether the experimenter controls ventilation or not; if he is interested in the overall effect, then the secondary effects arising from a changed ventilation are an important part of the response. However, if he wishes to describe the direct effect, he must control the level of ventilation.

A similar example concerns the effects of stimulation of the sympathetic nerves on the coronary vessels. When the sympathetic nerve supply to the heart is stimulated, coronary blood flow increases but again this is a secondary effect. The sympathetic nerves will increase the rate and force of contraction of the heart, this will increase the oxygen consumption of the heart and consequently coronary blood flow will be increased. If the effects of rate and force of contraction are eliminated, then stimulation of the sympathetic nerves results in a vasoconstriction and thus a reduction in coronary blood flow. Thus, again the initial direct effects are reversed by the indirect compensatory effects. This can be seen as an extreme example of the effects of homeostatic control mechanisms.

8.3 Modifying hypotheses

As a result of the outcomes of experiments, either a rejection or modification of the original hypothesis will be required or the existing one will have to be tested more stringently. Sometimes, these processes of modification and refinement result in hypotheses that were initially rather distinct beginning to merge.

Consider the models of Figure 8.1 where it might be postulated that '*Q*' stands for mental performance and '*P*' for body temperature. There is evidence that body temperature does affect mental performance, in accord with model A. However, there is further evidence that the state of CNS arousal affects both variables (this could be designated 'X' and could then be evidence in favour of model B). It would seem that a decision between the two models could be made if variable P alone (body temperature) were changed. Model A predicts a change in Q (mental performance) but model B does not. However, independent results indicate that changes in body temperature exert an effect not only upon mental performance but also upon general CNS arousal (X in model B). That is, an effect of temperature upon mental performance might be a direct one or an indirect one through, say, CNS arousal. The distinction between the two models becomes less clear-cut since the system seems to possess elements of both models.

Another example could be based upon the observation that salivary glands increase in size (hypertrophy) with increased use. One hypothesis ('trophic') could be that nerves controlling salivary glands release a chemical factor that causes the glands to grow; another ('metabolic') could be that the glands enlarge due to their increased secretory processes or metabolic rate; and the third ('electrical') could be that hypertrophy is due to the electrical changes produced across the cell membranes when the gland is active. In certain circumstances, it would be possible to distinguish between these hypotheses. Thus, the first (or trophic) hypothesis would seem most appropriate if a chemical substance could be extracted from nerves and it could be shown that this substance was released at the neuro-effector junction during stimulation and that, when applied directly to unstimulated salivary gland cells, it did not produce the electrical and metabolic effects associated with secretion but could reproduce exactly all the features of cell hypertrophy. The second (or 'metabolic') hypothesis would be favoured if a substance could be found that accumulated (or decreased in concentration) when the glands showed metabolic secretory activity and which, when injected into (or removed from) *unstimulated* gland cells, could produce cell hypertrophy. The third (or 'electrical') hypothesis would be favoured if a denervated cell were directly stimulated electrically and its size was determined wholly by the amount of electrical stimulation it received rather than by the amount of secretion or osmotic work that it performed.

However, when one considers:

(1) that nerves release chemical transmitters that are associated with electrical changes in gland cells;

(2) that electrical changes in gland cells are associated with metabolic changes in gland cells; and

(3) that secretory processes are associated with metabolic and electrical changes;

then it becomes clear that to determine the specific stimulus promoting hypertrophy would be very difficult. Firstly, there are a technical difficulties, since the stages between neural stimulation of the gland and its secretory activity are closely linked; in order to distinguish between the rival hypotheses, it would be necessary to change one process whilst keeping the others constant. It is also possible that none of the three hypotheses is appropriate. It may be that causing a glandular cell to secrete produces a group of parallel changes that interact — rather than act individually — to produce a number of effects including the control of gland size. If this is so, then a single stimulus to hypertrophy does not exist and some amalgamation of the initial hypotheses would be more appropriate; the difficulty here stems from the starting hypotheses which inappropriately divided the secretory process into a number of discrete components. Similar difficulties in interpretation would arise if the neurotransmitter were the factor causing hypertrophy in addition to effecting secretion or if, when the gland was secreting, much of the extra metabolic activity was used to regenerate the ionic gradients across the cell membranes that had been dissipated during secretion. In such cases, the distinctions between the hypotheses become blurred to the extent that they might be little more than semantic differences.

8.3.1 *Problems when hypotheses are modified*
Two further kinds of problem exist. The first is that as the tests of a hypothesis become more refined, a position is reached in which the technical requirements of the tests become a limitation. Thus a distinction between rival hypotheses postulating whether or not there was continuity between nerve and muscle at the nerve–muscle junction awaited development of the electron microscope with a sufficient resolving power; and whether or not the peak of the action potential resulted in a reversal of the membrane potential awaited the development of intracellular electrodes. This idea can be illustrated by considering the wealth of new information that has followed the introduction of new techniques — for examples, radioactive tracers, electronic recording devices, chemical microanalytical methods, radioimmunoassay and computers — and the promise for the future that exists for

recently developed techniques such as genetic engineering and mono-clonal antibodies. Equally, one can think of many physiological problems, a possible solution to which is limited by the inadequacy of our present instrumentation. If there is a general rule here it seems to be that the next technique will often be developed in another (often quite unrelated) branch of science — 'spin-off'. Before this happens, the experimenter must either wait or try and approach the problem by a different route. For example, techniques are not yet developed which will allow us to 'see' at all clearly the pores which have been postulated to exist across cell membranes. Accordingly, their presence has been sought indirectly either by attempting to isolate and study the three-dimensional structure of proteins that render membranes porous or by attempting to ascertain mathematically the size of pores that would account quantitatively for the permeability of membranes. It will be noticed that there is a link here between developing new experimental approaches and devising new hypotheses by juxtaposing previously disparate ideas (see Chapter 1).

The second problem is that the accumulation of experimental results and the continual refinement of a hypothesis that this produces can result in the development of an exceedingly complex model of a process. Of course, such a model might be correct but a distinction should be drawn between models based upon complex hypotheses and mechanisms and those based upon a simple underlying hypothesis and mechanism that is flexible enough and has sufficient potential to produce a whole variety of complex outcomes. For example, when different types of haemoglobin molecules are considered, they often differ functionally from each other in complex ways. However, this often seems to result from small differences in their amino acid sequences, as a result of which the same set of rules governing the folding of an amino acid chain into a three-dimensional structure results in a wide variety of molecular shapes and functional properties. To take a further very simple example, consider an array of four switches, each of which when taken singly could be either 'on' or 'off'. By contrast, when considered as a group, this array could show 16 different arrangements, from 'all on' to 'all off'. The important point is that the properties of this system are far more complex than those of its components. These ideas are sometimes expressed as the rule of parsimony or the principle of Occam's razor, the point being that one should always attempt the simplest hypothesis to explain the results.

8.4 **Mathematical and physical models**

As the process of refinement continues, the hypotheses take on a more quantitative nature. For example, having established that gas A increases ventilation (see Figure 6.10), the experimenter might wish to refine his hypothesis by investigating the quantitative relationship between the change in concentration of gas A and the increment over control ventilation rate that this produces. Such a hypothesis can be described in mathematical terms. For example, if he postulates that the relationship is a linear one, then the equation becomes:

$$\dot{V} = C + m \cdot [A]$$

where

$\dot{V}$ is the rate of ventilation

$[A]$ is the concentration of gas in the blood

C is the (control) rate of ventilation in the absence of the gas, and

m is a constant; it is the gradient of the line relating $\dot{V}$ to $[A]$ and could be defined as the sensitivity of the whole body to the effects of gas A.

Thus, what the experimenter is doing is producing a mathematical model of how the physiological system would behave. (Such mathematical models should not be confused with the biological models discussed in Chapter 2.)

Such a relationship could then be investigated by measuring $\dot{V}$ after administration of different amounts of the gas and so raising the concentration of gas A in the blood. One way to analyse the results would be to estimate the values of C and m by linear regression analysis. This involves the use of Pearson's product–moment correlation coefficient, r; the analysis would also enable the confidence intervals for both C and m to be estimated. Note that, in performing such a statistical analysis upon the results, the *assumption* is made that there is a linear relationship (see also Fig. 6.8 and the associated discussion). If the correlation coefficient is not significantly different from zero, or if its value is significantly less than zero, then the assumptions on which the mathematical model are based would be disproved. (In fact, if the correlation coefficient is not significantly different from zero, then any line describing the data is valueless.) However, if the correlation coefficient is significantly greater than zero, the possibility remains that the results would be better described by some other mathematical function, by a logarithmic relationship, for instance. (The preference of one mathematical model to another is a highly specialized topic and beyond the scope of this book.)

In spite of these reservations, the strength of a mathematical model is that it enables a hypothesis to be described in quantitative terms. Mathematical models can become exceedingly complex especially in cases where the homeostatic systems of whole animals are being described cybernetically. In such cases, considerable numbers of negative feedback loops can be modelled, often with many of them interacting with each other. Sometimes, the result is so complex that the model has to be tested on a computer.

In such tests, the model is required to predict the effects of a particular disturbance or set of circumstances. Thus, if the model describes the process of thermoregulation, it can be tested by investigating the effects of a particular thermal environment or by using a drug to change the activity of the hypothalamus. The adequacy of such a model is generally judged by the accuracy of its predictions compared with actual observations. If the model is successful, then its usefulness is that it can be used to predict the outcome of circumstances that might be dangerous in real life or expensive to perform. Unfortunately, this usefulness is negated to an unknown extent by the fact that the nature of the experiments is such that extrapolations are having to be made from the model to the living organism.

In addition, in all such models, of whatever degree of complexity, the explanation of what, in physiological or biochemical terms, is meant by the terms in the equations might remain a problem. For instance, is the sensitivity of the system to gas A (the term m in the equation above) a reflection of properties of the chemoreceptor, of the CNS or of the motor units controlling ventilation — or of some combination of all of these factors? Clearly, further experiments would have to be performed to establish the answer to this question.

This problem of determining the physiological or biochemical correlates of a model can also be particularly difficult when physical models of living systems are being described. Consider the following models:

(1) The passage of water-soluble molecules across capillaries and cell membranes can be modelled in terms of movement of particles through cylindrical, water-filled channels of a certain dimension.
(2) The electrical changes across the membranes of excitable tissue cells occurring during the generation of an action potential can be

described in terms of 'equivalent electrical circuits' which make use of the known resistance and capacitance properties of the membrane or (and this is closely related to (1) above) can be described in terms of the time-courses of opening and closing of 'gated' channels.

(3) The production of a large concentration gradient rising from the cortex towards the papilla of the kidney can be described in terms of a countercurrent multiplier system.

(4) The mechanical properties of whole muscle can be described in terms of a contractile element (sliding filament activity) and series and parallel elastic components.

(5) The rhythmic alternation between inspiration and expiration and the possession of clock-like activity by the hypothalamus of mammals can be modelled by groups of neurones interacting with each other in accord with rules derived from an understanding of the way coupled oscillators (whether electrical or mechanical) are known to work.

All such models not only allow a quantitative description of biological processes to be given but also enable concepts learned from cybernetics, physics and electronics — the properties of electrical circuits, counter-current multipliers and oscillators, for instance — to be incorporated into biological models. However, as with mathematical models, their disadvantages must always be borne in mind. They are always mathematical or physical descriptions of biological processes or analogues of the biological world. Now, even though we do not imply anything special about biological material in a philosophical sense, we must make it clear that it would be naive to expect to see actual springs in muscle, resistors and capacitors in a membrane or rigid cylinders inserted across capillaries. Instead, the properties of the biological tissue are such that they act *as if* they possessed these structures and the exact means by which such properties arise in biological systems is not addressed by such models. (This is why, for example, the term '*equivalent* pore radius' is used in connection with model 1, above.)

Unfortunately, this lack of correspondence between the components in a model and the structures in biological material can lead to difficulties of interpretation. Thus, it is quite possible to alter a physical or mathematical constant in order to make the model agree more closely with the experimental results, but the exact meaning of such a modification in biological terms might not be clear. To describe the rhythmic changes produced by the respiratory regions of the medulla and the pons in terms of structures which show self-re-excitation and reciprocal inhibition does little to identify the biological structures involved in this process; and models able to describe the action potential in a nerve do not address themselves to the problems of identifying how the variable resistances (channels and gates) that are believed to be involved actually work. There is a further difficulty in that the immense versatility of physical and mathematical modelling can lead to models based on different premises producing indistinguishable results; in the absence of biological structures that correspond to the different parameters of each model, a distinction between the models as a description of a biological process can be most difficult to make.

8.5 Some conclusions

In this part of the book we have discussed some aspects of the scientific method as it applies to physiology. We have concentrated upon the need for the experimenter to take care at all stages of an experiment; at the initial planning stage, during the execution of the experiments and in the last stages of drawing conclusions. Thus, there is a need for the utmost rigour, self-discipline and integrity on the part of the experimenter.

But there is another side. The successful experimenter not only shows all of these attributes but also shows characteristics which, though they are more nebulous, are no less important. We have already considered how advances may be made when individuals take notice of, and realize the significance of, 'anomalous results' or piece together information from differing sources. Such aspects of scientific work demand that an investigator uses his ingenuity and imagination. Those who work in research accept the necessity for boring, repetitive work and are rewarded by the occasional 'moment of truth', 'breakthrough' or 'spark' that comes to them.

Research is a tremendous intellectual challenge that demands much of its experimenters but in return can give considerable satisfaction. There remain in physiology many challenges to be met, many opportunities to test researchers' ingenuity and skill — and to provide intellectual satisfaction and excitement. We hope that this book might help those embarking on a research career to gain these rewards as a result of a high quality of experimentation.

Part II

Introduction

The aim of this part of the book is to illustrate the principles described in the first part by means of abstracts based loosely on published papers. Each is in three parts: the abstract itself; some questions related to it; and some guidelines to the answers.

The papers were initially chosen because of the topics they cover. Although students are unlikely to have studied these topics in any detail, they should be readily understood. We have freely changed the text, results, figures, etc. of the original papers so that the resulting abstract is more suitable for our own requirements. That is, the finished abstract need no longer and probably does not represent the original paper, nor need it represent the views of the paper's authors.

The initial abstracts deal with *in vitro* work. Later abstracts deal with whole animal preparations and the final paper deals with a study in man.

The questions direct attention to many aspects of the scientific method as they have been put forward in Part I. There are some areas which we tend not to cover fully even though, as indicated in Part I, they are of great importance, for example details of the measurement. We would not expect students to be conversant with the detailed characteristics of a particular instrument. In these abstracts, as must normally be the case with any scientific report, it is assumed that the apparatus was suitable for the task it was used for, that it was maintained and calibrated correctly and that the experimenters used it properly.

The guidelines are not intended as 'model' answers! Instead, they represent one approach which might direct the student along an appropriate path. At times we have elaborated upon certain aspects of our guidelines where we feel that the issues involved are sufficiently important. Obviously, no student would reasonably be expected to think of all these points. Instead, we would stress that many other answers are possible and we encourage students to think of them. Indeed, the whole aim of the book is that the reader should decide for himself the validity, or otherwise, of the abstract and the guidelines.

Finally, we point out that, even though the abstracts can be considered in isolation, reading them in the order given has one advantage: if a problem arises in more than one abstract, it is covered in most detail on the first occasion; subsequently only extra points are raised or particular ones stressed.

Abstract A

The effect of plasma albumin concentration on albumin synthesis by the liver

Introduction

The concentration of albumin in plasma is normally very constant and is believed to be regulated by changes in the rate of synthesis of albumin by the liver. Several factors have been suggested as mediators of these changes including the concentration of albumin in the plasma and the colloid osmotic pressure (COP) of the plasma.

Previous studies using an isolated perfused liver preparation have shown that the rate of synthesis of albumin by the liver is inversely proportional to the concentration of albumin in the perfusate. However, it is not possible from these experiments to conclude whether it is the concentration of albumin in the plasma *per se* or its colloid osmotic pressure which regulates the rate of synthesis. The purpose of the present investigation is to differentiate between these two factors in their effect on the rate of albumin synthesis.

Methods

Experiments were carried out on female hamsters weighing about 150 g and anaesthetized with pentobarbitone ($70 \, mg \cdot kg^{-1}$ intraperitoneally). Prior to the experiment, they were fed *ad libitum* on a normal diet. The portal vein was cannulated and then the inferior vena cava was cut distally in the abdomen and perfusion of the liver via the portal vein was started as quickly as possible. This protocol reduced the time during which the liver was not perfused to about 1 or 2 min. Blood vessels and connecting tissue surrounding the liver were then sectioned and the liver transferred to paraffin oil. Before each experiment the liver was perfused for 30 min to allow the temperature and flow rates to stabilize.

The livers were perfused with two solutions; these were modified so that eight types of perfusion experiment were performed. The first solution was free of plasma and consisted of washed bovine erythrocytes suspended in Ringer's solution which contained normal concentrations of amino acids and glucose. Bovine serum albumin was then added to this solution so that the final concentrations of albumin were 40 (group I), 30 (group II) and 15 (group III) mg $\cdot$ ml^{-1}. The haematocrit of these solutions ranged between 24 and 26 per cent. The second solution was composed of erythrocytes (as above) mixed with bovine plasma and Ringer's solution so that the ratios (by volume) were 2:4:2, 2:2:4 and 2:1:5 respectively. The concentration of protein in these three solutions were 48 (group IV), 24 (group V) and 12 (group VI) mg $\cdot$ ml^{-1}.

In order to change the colloid osmotic pressure without a simultaneous change in the albumin concentration, bovine gammaglobulin was added to diluted plasma containing 24 (group VII) and 12 (group VIII) mg protein $\cdot$ ml^{-1}.

The rates of synthesis of albumin were determined by measuring the increase in hamster albumin in the perfusate leaving the liver. Perfusion volume was measured by isotope dilution using [^{125}I]-albumin. O_2-uptake was measured by estimation of the O_2 concentrations in the perfusate entering and leaving the liver.

In two experiments with a medium as in group II and with normal rates of albumin synthesis, cycloheximide (1 mM) which blocks the synthesis of albumin was added after 90 min of perfusion.

Results

In all the experiments, apart from those in which the livers were perfused with a plasma-free solution containing 15 mg $\cdot$ ml^{-1} of serum bovine albumin (group III), average oxygen uptake varied only slightly between different experiments starting at an average value of 2.5 (μmol $\cdot$ min^{-1} $\cdot$ g liver wet weight^{-1}) and decreasing to, on average, 2.1 at the end of the experiment. However, in the studies using group III, the oxygen uptake decreased during the course of the experiment. In the first 2 h, the uptake averaged 1.9 and decreased in the last hour to 1.5 μmol $\cdot$ min^{-1} $\cdot$ g liver wet weight^{-1}.

Table A.1 shows the results when a plasma-free medium was used and results from the experiments with bovine plasma are given in Table A.2.

After giving cycloheximide, the rates of synthesis in the period from 150 to 210 min were only 4 and 7 per cent of the average rates of synthesis before addition.

Table A.1 Rates of albumin synthesis in hamster liver (mg.h^{-1}.100 g body weight^{-1}) perfused with one of three plasma-free media. The results are given as mean ± SE

Time (min)	Group I (n = 5) Albumin: 40mg·ml^{-1} Colloid osmotic pressure (COP): 20.3 cmH$_2$O	Group II (n = 8) Albumin: 30 mg·ml^{-1} Colloid osmotic pressure (COP): 15.2 cmH$_2$O	Group III (n = 6) Albumin: 15 mg·ml^{-1} Colloid osmotic pressure (COP): 7.6 cmH$_2$O
30–90	1.6 ± 0.3	2.1 ± 0.1	1.8 ± 0.2
90–150	1.2 ± 0.3	2.4 ± 0.2	1.8 ± 0.3
150–210	1.0 ± 0.1	1.5 ± 0.2	1.1 ± 0.2
Mean 30–150	1.4 ± 0.2	2.2 ± 0.1	1.8 ± 0.2
t tests	I v. II: P < 0.05 I v. III: n.s.	II v. III: P < 0.05	

n.s., not significant at the 5 per cent level

Table A.2 Rate of albumin synthesis in hamster liver (mg.h^{-1}.100 g body weight^{-1}) perfused with one of five perfusates consisting of a mixture of bovine erythrocytes, a Ringer's solution and bovine plasma

Time (min)	Group IV (n = 7) Plasma protein: 48 mg·ml^{-1} Albumin: 20 mg·ml^{-1} Colloid osmotic pressure (COP): 17.5 cmH$_2$O	Group V (n = 6) Plasma protein: 24 mg·ml^{-1} Albumin: 10 mg·ml^{-1} Colloid osmotic pressure (COP): 8.7 cmH$_2$O	Group VI (n = 6) Plasma protein: 12 mg·ml^{-1} Albumin: 5 mg·ml^{-1} Colloid osmotic pressure (COP): 4.4 cmH$_2$O	Group VII (n = 5) Plasma protein: 56 mg·ml^{-1} Albumin: 10 mg·ml^{-1} Colloid osmotic pressure (COP): 17.1 cmH$_2$O	Group VIII (n = 7) Plasma protein: 30 mg·ml^{-1} Albumin: 5 mg·ml^{-1} Colloid osmotic pressure (COP): 9.0 cmH$_2$O
30–90	1.4 ± 0.2	1.9 ± 0.2	2.5 ± 0.2	1.4 ± 0.2	2.1 ± 0.2
90–150	1.6 ± 0.1	2.1 ± 0.2	2.9 ± 0.3	1.1 ± 0.2	2.0 ± 0.2
150–210	1.3 ± 0.2	1.9 ± 0.4	2.4 ± 0.2	1.6 ± 0.2	2.2 ± 0.2
Mean (30–210)	1.4 ± 0.1	2.0 ± 0.1	2.6 ± 0.1	1.4 ± 0.1	2.1 ± 0.1
t tests	IV v. V: $P < 0.01$ IV v. VI: $P < 0.001$ IV v. VII: n.s. IV v. VIII: $P < 0.01$	V v. VI: $P < 0.01$ V v. VII: $P < 0.01$ V v. VIII: n.s.	VI v. VII: $P < 0.01$ VI v. VIII: $P < 0.01$	VII v. VIII: $P < 0.01$	

Discussion

Using a synthetic medium containing bovine erythrocytes and diluted or undiluted bovine plasma, it was possible to perfuse the liver for up to 5 h with only a slight decrease in albumin synthesis (see Table A.2) and oxygen uptake. However, the experiments with plasma-free medium (see Table A.1) gave satisfactory results only when an albumin concentration of more than 15 mg $\cdot$ ml^{-1} was used. Thus, with group III, it was not possible to continue the perfusion for more than about 2½ h. The reason for this is uncertain; however, diluted plasma with only 12 mg protein $\cdot$ ml^{-1} (group VI) gave a stable perfusion with a satisfactory flow rate and oxygen uptake.

In these experiments, there was a significant negative correlation between colloid osmotic pressure and the rate of albumin synthesis. It is therefore concluded that the rate of synthesis of albumin is regulated through the colloid osmotic pressure. The albumin concentration plays a role in regulation only through its contribution to the colloid osmotic pressure.

Questions

A.1 How physiologically viable do you consider the experimental preparation to be?

A.2 *(a)* Explain how the rates of albumin synthesis were calculated. What assumptions are made in those calculations?
(b) Comment upon presentation of rates of albumin synthesis as 'mg $\cdot$ h^{-1} $\cdot$ 100 g body weight^{-1}'.

A.3 What comments have you to offer with regard to the general protocol and the choice of perfusion media?

A.4 From a statistical viewpoint, how well do the results support the view that the rate of albumin synthesis is a function of the colloid osmotic pressure of the medium rather than its albumin concentration *per se*?

A.5 From a physiological viewpoint, how reasonable do you consider the hypothesis — that the synthesis of plasma albumin is controlled wholly by colloidal osmotic pressure — to be?

Suggested Answers

A.1. The aim of the present experiment is to distinguish between the roles of colloidal osmotic pressure and concentration *per se* in the control of albumin synthesis by the liver. To have investigated the effect of protein infusions upon liver albumin synthesis in a whole animal preparation would have been very problematical. This is because the amount of the infused protein reaching the liver and, even more so, the rate of synthesis by the liver would have been almost impossible to estimate. A system in which these two factors — the stimulus and the response — could be measured accurately is needed. *In vitro* studies using, for example, liver slices are also limited in their usefulness since the natural means of delivering the stimulus to the liver, the blood supply, would have been lost.

Perfusion studies of a single organ are potentially so useful because, properly performed, these experiments will allow very considerable control over the stimulus and comparative ease in measuring the output from the organ.

The liver poses special problems mainly by virtue of its blood supply which consists not only of a conventional arterial supply but also of a portal input from the gut which is at a much lower pressure. Normally, the liver receives about two-thirds of its oxygen and nutrients from the portal system. Perfusion by the liver through the portal system must attempt to substitute for both blood supplies. Accordingly, perfusion parameters (pressure, flow, etc.) and the composition of the perfusate (O_2, glucose, etc.) are likely to be difficult to adjust in a completely satisfactory manner as a result of attempting to replace a dual blood supply.

Another problem exists when attempts are made to study an organ the size of the liver in isolation. If it is removed from the support it normally receives *in situ* it will tend to become compressed under its own weight. This will hinder the perfusion of parts of the tissue. Immersion in oil will oppose this as would immersion in (say) a Ringer's solution. The oil will provide more of a barrier to water and salt loss than will aqueous environments. How closely either paraffin oil or a Ringer's solution will mimic the mesentery is unclear.

In the face of these potential difficulties, the efficacy of such a perfusion system must ultimately be judged by the ability of the liver to perform normally. The abstract gives evidence that the preparation is fairly *stable* (but not necessarily *normal*) with respect to protein synthesis and oxygen consumption (for all groups except III). Note also that subnormal albumin levels *increase* albumin production; this suggests that this function of the liver has been retained in spite of the abnormal composition of the perfusate. The values for oxygen consumption could be compared with values *in vivo* obtained from this species. Further assessment of liver function could be by its ability to secrete bile normally or by standard liver function tests, for example, the secretion into bile of bromosulphthalein.

There is evidence from the oxygen consumption values that experiments upon group III were upon a deteriorating preparation and so the results from this group should be treated with extreme caution. The cause of such deterioration is unknown; it did not occur in other groups with lower albumin concentration (V–VIII), lower COP (VI) or with no plasma (I, II).

A.2 (a). A standard method of measuring the amount of albumin secreted by the liver would be given by

$$S_{\text{T-I}} = V_{\text{T}} \cdot A_{\text{T}} - V_{\text{I}} \cdot A_{\text{I}} \tag{1}$$

where

$S_{\text{T-I}}$ = amount synthesized between time T and initial time, I
$V_{\text{T}}, V_{\text{I}}$ = volume of perfusate at times T, I
$A_{\text{T}}, A_{\text{I}}$ = concentration of albumin in perfusate at times T, I.

The volume of perfusate, V, is given by

$$V = \frac{VL \cdot CI}{CP} \tag{2}$$

where

VL is volume of labelled albumin injected
CI is radioactivity (counts $\cdot$ min^{-1}) of injected albumin
CP is radioactivity of perfusate.

This method assumes that the labelled albumin is not lost by liver metabolism, that the albumin has equilibrated throughout the perfusate and that the assay for albumin is specific for this class of molecule. Though the results are interpreted as a synthesis of protein rather than a release of stored protein, the method would assess the increment in 'protein pool' regardless of its source.

The experiments with cycloheximide support the assumption that protein synthesis rather than release of stored protein is involved. Thus the text indicates that, after this inhibitor had been administered at 90 min, protein synthesis between 150 and 210 min was 4–7 per cent of the value beforehand (30–90 min). This indicates values of about 0.08 and 0.15 mg $\cdot$ h^{-1} $\cdot$ 100 g body weight^{-1}. (The normal value at this time from the eight experiments in group II (see Table A.1) is

1.5 ± 0.2 mg $\cdot$ h^{-1} $\cdot$ 100 g body weight^{-1}.) Even though no wholly satisfactory statistical analysis can be performed upon these data, the large difference after cycloheximide addition is obvious and, in practice, few researchers would consider that a statistical validation was necessary. However, possible statistical approaches are:

(1) Treat data as two groups ($n = 8$ and 2) and then compare the two samples by t test.

(2) Estimate the population mean and standard deviation from the control sample of eight values (1.5, 0.6 respectively) and then for each cycloheximide experiment estimate t by equation:

$$t = \frac{\text{population mean} - \text{cycloheximide value}}{\text{SD}}$$

The number of degrees of freedom is 7. The t distribution will enable an estimate to be made of the chance of finding a value as extreme as (or more extreme than) the observed result assuming it came from the same population as the control sample. It would seem reasonable to use a one-tailed test since it is whether a value *as low as* the observed result is being considered.

Notice that calculating the rate of protein synthesis in the above manner is unnecessarily complex. Thus, since the volume of perfusate can be controlled at a known value, increases in concentration of albumin will reflect changes in amount present, that is, synthesis (or release). Therefore, equation (1) becomes

$$S_{\text{T-I}} = V(A_{\text{T}} - A_{\text{I}})$$

where V is the (constant) voluem of the perfusion system. As a third method, note that the specific activity of the albumin (counts $\cdot$ min^{-1} $\cdot$ g. albumin^{-1}) will decline during the course of the experiment due to synthesis by the liver of non-radioactive albumin. This calculation assumes that the labelled molecules are not metabolized by the liver, as did the first method of calculating synthesis.

A.2 (b). When results are presented in this form, namely as a ratio, a raised value can be due to a raised rate of synthesis and/or a lower body weight. In the present experiments, exact body weights are not given: if the different groups were matched for weight (or at least there was a random distribution of animals' weights between the groups) then presenting the results this way will not cause difficulties. However, if a group of animals happens by chance to be heavier than average, then a spuriously low rate of albumin synthesis would be calculated for this group. Another problem is whether the results should be expressed in terms of body weight. Certainly the results become easier to compare (and the variance within a group will be less) if albumin synthesis is directly related to body weight, but expressing the results in terms of another variable (liver weight?, blood volume?, plasma protein pool?) might be more justified physiologically, even if harder to do! In practice, the problem is ameliorated by the fact that all these variables correlate positively with body weight.

A.3. Theoretically, the experimenter should aim for a protocol which uses different perfusion media in random order (including a normal 'control' medium). However, the time that would have been required to do this, coupled with the possible inability to maintain the perfusion system satisfactorily for more than 5 h, thwarted this aim. Even so, the tables indicate that rates of synthesis were fairly stable over the period ½–3½ h — at least with the media containing bovine plasma — and so the protocol could have consisted of perfusing the liver with two different media and making an hour-long estimate of protein synthesis with each. However, if this were done then studies would have to be carried out to check that changes in the rate of albumin synthesis were apparent in this time. It may be too short a period. One medium could be a control medium (hamster plasma) and the second a randomly chosen 'test' medium; the sequence 'test/control' should be randomly ordered.

The presence in each experiment of a control perfusate would have enabled the viability of each preparation to be assessed by measuring protein synthesis and oxygen consumption (see question A.1). It would also have improved the type of calculations that could have been made (see later).

The main problem with the protocol is the choice of perfusion media. This problem is a compound one:

(i) *The use of two perfusion media.* This decision necessarily raises difficulties when results using the different media are compared. The use of a plasma-free medium to which albumin is added appears to be a more direct test of the original hypothesis. However, it suffers from two disadvantages. The first, which has already been discussed, is that there is evidence that, at least with low concentrations, the perfusion medium is inadequate to maintain the liver preparation in a 'healthy' condition. The second is much more fundamental. Since the aim of the experiments was to distinguish the effect of protein concentration itself from the colloidal osmotic pressure it produces, the perfusion medium is useless since only the albumin produces the colloidal osmotic pressure. That is, you cannot change one without the other changing in *exactly the same way*! The only role of such experiments would appear to be to confirm the previous results cited in the Introduction. Such a confirmation of earlier studies is often included in an experimental study.

(ii) *The distinction between albumin concentration and the colloidal osmotic pressure it produces.* The second group of perfusion media is slightly better in that it enables colloidal osmotic pressure to be changed independently of albumin concentration. However, this does not seem to have been done methodically. One method would have been to add different amounts of gammaglobulin to the control plasma; this would increase COP independent of albumin concentration. One could then further investigate the effect of other large molecules, for example, dextrans, fibrinogen, in substituting for gammaglobulin. This method has the disadvantage that a reduction in COP cannot be produced. It might be better, therefore, for the control medium to be plasma to which had been added (say) some dextrans so that its COP was higher than normal; then the effect of falls in COP as well as rises (compared to this 'control') could be measured.

A second method, which complements the first, is to keep the total COP constant and then to replace albumin with gammaglobulins (or other large molecules). Further, if the control medium is that with the raised COP (see above) then the effect of increasing albumin concentration can be investigated by removing an osmotically equivalent amount of other material.

(iii) *The use of bovine blood.* The position with respect to blood transfusion is far simpler in the hamster than in primates. Problems due to heterologous blood do not seem to arise. Further, if more hamster blood were required, then a series of donor animals would be killed and bled. In this case, the experimenter is fortunate in that a potential problem does not materialize.

A.4. The hypothesis being tested is that there is a negative correlation between the COP exerted by the albumin in the perfusate and the rate of synthesis of albumin by the liver. An important rider to this hypothesis is that there should not be a negative correlation between the concentration of albumin and the rate of synthesis. Accordingly, the data from Table A.1 (groups I to III) are of no use since it is albumin alone that is responsible for COP (in other words, there are bound to be identical correlations between protein synthesis and albumin concentration and protein synthesis and COP). There are results from the first perfusion medium (groups I−III) which are *against* the hypothesis. Thus, group III has a lower rate of albumin synthesis than group II even though the former group has less albumin and a lower oncotic pressure. Arguably, the results from group III can be ignored because there is *independent* evidence that the livers were in an unsatisfactory condition. In such circumstances, one must rely completely upon the integrity of the experimenters.

Pearson's product−moment correlation, coefficient, r, can be calculated for the results from groups IV−VIII of Table A.2. Values from 150−210 min give some indication of declining synthesis so, in the following comments, only values betwen 30−150 min have been used. When protein synthesis and COP are compared then the correlation coefficient equals -0.95, a value sufficiently large to allow the null hypothesis (that there is no correlation) to be rejected at the 5 per cent level. However, there are two difficulties with this approach. The first is technical; if average values are used, then the correlation coefficient will be further from zero than would be the case if the individual values were used. Thus there is an increased chance when average values are being used of making a type I error (falsely rejecting the null hypothesis). The second difficulty is that a similar analysis of the correlation between albumin concentration and protein synthesis might give a similar result. In other words, one cannot reject the alternative hypothesis on the basis of such results.

However, there is an alternative analysis of the data of Table A.2. Thus groups V and VII and groups VI and VIII form two pairs. The members of each pair have identical albumin concentrations but differ in COP. The results of t tests indicate significant differences between the member of each pair. That is, COP affects protein synthesis when albumin concentrations are unchanged. By contrast groups IV and VII and groups V and VIII form two other pairs: members of each pair have similar COP but different albumin concentrations. The results of t tests here indicate a lack of significant differences between members of the pair. That is, albumin concentration does not affect protein synthesis significantly if COP is unchanged.

These comparisons, not the correlations, provide the best evidence that it is the COP, not albumin concentration, that is changing protein synthesis.

Note that, had the protocol enabled more than one perfusion medium to have been used on the same preparation (see question A.3) then more powerful paired statistical tests could be used.

A.5. One of the roles of plasma proteins, and one in which albumin exerts a most important effect, is to maintain plasma–interstitial fluid balance; this requires an accurate control of plasma COP. However, this will not apply in the liver since the capillaries in this organ are discontinuous; rather, any COP effects will be exerted across the cell membranes of the liver cells themselves.

The concept that the COP of plasma is controlled through synthesis of albumin by the liver in response to changing values of COP is an attractive one, another example of a homeostatic mechanism working by negative feedback. Even so, some problems are raised by this concept concerning the specificities of stimulus and response. The present work has used gammaglobulins to suppress albumin synthesis: do other proteins — including albumin itself — or high molecular weight substances have the same effect? If the suppression were wholly through changes in COP, then one would predict that they would. Further, clinical conditions in which gammaglobulin concentrations were abnormal would have predictable effects upon the synthesis of albumin by the liver. An additional problem is that the 'albumin' fraction of plasma proteins is heterogeneous: how specific for different albumin fractions was the assay that was used and was the synthesis of all fractions stimulated equally?

Finally, one notes that the changes in total protein and albumin concentrations and COP that were used ($12-56$ mg $\cdot$ ml^{-1}, $5-40$ mg $\cdot$ ml^{-1} and $4.4-20.3$ cmH$_2$O respectively) were unphysiologically wide; to what extent this produces changes in liver function that do not represent the normal mechanisms of control (rather than an accentuation of these) is not known. When looking at the ranges of values used, it is also important to relate these to those found in the species being studied. For example, in the present investigation, the normal value for the concentration of total plasma proteins in the hamster is 45 mg $\cdot$ ml^{-1} compared to the value of 70 mg $\cdot$ ml^{-1} found in man.

The effect of increased salt loading on salt glands

Introduction

It has been known for some time that, in birds that live by the sea or in the estuaries of rivers, the size of the salt glands is related to the salinity of the birds' drinking water. It has been shown that when the salinity of the water is increased there is a hypertrophy, that is, an increase in the *size* of the cells in the glands. However, there is still some disagreement as to whether there is hyperplasia, that is, an increase in the *number* of cells in the glands, as would be indicated by an increase in the total amount of DNA present in the glands. There is also doubt as to the time-course of any compensatory changes.

The present investigation was carried out to determine, first, whether there is any hyperplasia of the salt glands at any time after ducks start to drink salt water. In a further series of experiments, the effect of removel of one salt gland on the growth of the remaining gland was investigated in geese drinking fresh water or salt water.

Methods

Experiments were carried out on adult domestic ducks and geese that prior to the experiment had been given fresh water to drink. The experimental group of animals were given 0.3 M NaCl salt water to drink and this was their sole source of drinking water. The presence of hyperplasia was assessed by measuring the rate at which labelled thymidine was incorporated into DNA. Studies were made of the rate of incorporation of labelled thymidine both *in vivo* and *in vitro*.

Experiment 1: Incorporation of labelled thymidine in vivo
Five groups, each containing five ducks, were studied. The control group had received only fresh water to drink and the remaining four groups

were given salt water to drink for 1, 2, 7 or 14 days; 3 h before the birds were killed, the labelled thymidine (60 μCi) was injected intravenously. The salt glands were quickly removed after killing the birds and were stored at $-20\,°$C.

Experiment 2: Incorporation of labelled thymidine in vitro

Three groups, each containing four ducks, were studied. The experimental groups were given salt water for 2 or 4 days to drink and the control group was given fresh water for 4 days. All groups were then killed and slices of the salt glands incubated for 3 h in a medium containing labelled thymidine.

Experiment 3: Effects of removal of one salt gland

In these experiments 20 geese were used; 10 of these were not operated upon and these formed the control group, and in the other 10 (the experimental group) the left salt gland was removed under anaesthesia. The animals were then allowed to recover for 10 days. Five animals from the control group and five from the experimental were then given fresh water to drink and the remaining five from each group were given salt water to drink for 14 days. At the end of the 14 days, all 20 birds were killed and the glands removed.

Chemical analyses

Estimates were made of the DNA, RNA and protein present in the salt glands. The radioactivity present in the DNA was determined and expressed as counts/min/μg of DNA (c $\cdot$ min^{-1} $\cdot$ μg^{-1}).

Results

The effects of giving ducks salt water to drink are shown in Figure B.1, which gives the results from the *in vivo* studies, and Table B.1, which gives the results from the *in vitro* studies.

The effects of removal of one salt-gland are shown in Table B.2.

Discussion

The results (see Figure B.1, Table B.1) clearly show there is an increase in the number of cells in the salt glands when ducks first drink salt water. However, it is not known whether this increase in the number of cells is due only to an increase in the rate of cell division rather than a decrease in cell loss.

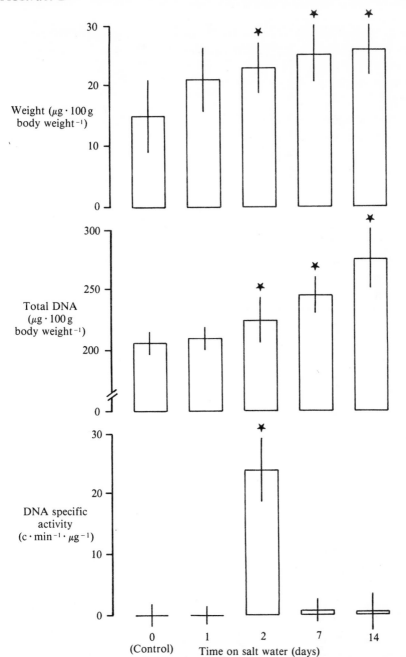

Fig. B.1 Changes in *in vivo* weight, total DNA and DNA specific activity in the salt glands of ducks given salt water; each group consisted of five ducks. Mean ± SE. *, $P < 0.05$, compared with control ducks given fresh water.

Table B.1 Weight and *in vitro* incorporation of labelled thymidine into DNA in the salt glands of ducks given fresh water or salt water. Mean ± SE

	Fresh water	Salt water	
Salt water given for (days)	–	2	4
Number of ducks	4	4	4
Body weight (kg)	3.13 ± 0.25	3.05 ± 0.19	2.86 ± 0.10
Salt gland weight (mg · 100 g body weight^{-1})	11.9 ± 1.5	25.5 ± 0.6	28.4 ± 3.9
		$P < 0.001$	$P < 0.01$
DNA specific activity (c · min^{-1} · μg^{-1})	4.1 ± 0.5	49.0 ± 13.9	43.0 ± 13.5
		$P < 0.05$	$P < 0.05$

P values derived by comparisons with group given fresh water

Table B.2 Weight, DNA, RNA and protein contents of right salt glands in which the left salt gland had been removed (operated) and in control (intact) geese. Fresh or salt water was given.

	Fresh water		Salt water	
	Intact	Operated	Intact	Operated
Time from operation to killing (days)	–	24	–	24
Salt water given for (days)	–	–	14	14
Number of geese	5	5	5	5
Body weight (kg)	4.72 ± 0.17	4.43 ± 0.23	5.06 ± 0.29	4.75 ± 0.38
Right salt gland:				
Weight (mg · 100 g body weight^{-1})	5.8 ± 0.4 A: n.s.^{d}	5.7 ± 0.4 B: n.s.^{a}	11.7 ± 0.5 C: $P < 0.001^a$ D: n.s.^{d}	15.3 ± 1.4 E: $P < 0.001^b$ F: $P < 0.05^c$
Total DNA (μg · single gland^{-1})	235 ± 20	232 ± 18 n.s.^{a}	337 ± 40 $P < 0.05^a$	338 ± 36 $P < 0.05^b$
Total RNA (μg · single gland^{-1})	103 ± 12	112 ± 14 n.s.^{a}	178 ± 16 $P < 0.01^a$	209 ± 25 $P < 0.05^b$ n.s.^{c}
Total protein (μg · single gland^{-1})	250 ± 10	281 ± 40 n.s.^{a}	471 ± 40 $P < 0.001^a$	714 ± 97 $P < 0.001^b$ $P < 0.05^c$

P values derived by comparison with: [a] intact birds given fresh water; [b] operated birds given fresh water; [c] intact birds given salt water; [d] left gland in same group; n.s. = not significant. For letters A–F, see question B.4.

The other experiments (see Table B.2) show that removal of one salt gland does not produce any compensatory growth in the remaining gland if the birds are given fresh water to drink. In birds given salt water to drink there was some compensatory growth but this was due to an increase in the size rather than the number of cells. These results support the view that hypertrophy in the salt gland is induced by an increase in functional demand instead of the removal of the other salt gland.

Questions

B.1 Have you any comments with regard to the choice of animals for these experiments?

B.2 What are the advantages of expressing:

(a) salt-gland weight as mg · 100 g body weight^{-1} (see Tables B.1 and B.2 and Figure B.1);

and (b) DNA, RNA and protein content as 'μg · single gland^{-1}' (Table B.2)?

B.3(a) How can the data of Figure B.1 be used to deduce if hypertrophy and/or hyperplasia has occurred?

(b) Do these data show the time-course of any changes?

(c) Do the data from Table B.1 add further information on these points?

B.4 Explain why the six statistical tests (labelled A−F in Table B.2) have been performed.

B.5 Do you agree with the last sentence of the discussion?

B.6 (a) Outline how you would investigate if a decreased rate of cell loss contributed to the present findings (see discussion).

(b) Of what use would a histological investigation be?

Suggested answers

B.1. As implied by the introduction, not all birds possess salt-secreting glands and those without such a gland would, of course, have been inappropriate. The domestic species used in these experiments can tolerate brackish water and, accordingly, their glands are not as well developed as certain sea-dwelling birds. The experiments investigated the effect upon their glands of exposing them to salt water for the first time. Note that, in this context, the strength of salt water used in these experiments (0.3M NaCl) could be regarded as 'physiological'. An advantage of the protocol is that, by using animals that had not drunk salt water before, the baseline or control conditions could be accurately described. An alternative approach would have been to investigate the amount of atrophy and aplasia occurring in the glands of marine birds which were exposed to diluted salt water. The choice of one species for some of the experiments and a second species for others may lead to difficulties in interpretation if comparisons are made between data obtained in the two groups of experiments.

B.2 (a). An adaptive change of salt glands would be expected to be specific rather than involve a generalized increase throughout the body. Since animals (and their glands) grow during the course of the experiment, account of this must be taken. This is achieved when results are expressed as 'weight of salt glands per unit body weight'. An alternative way would be to express them in terms of the weight of some other organ (heart, liver) which would not be expected to change with salt water drinking. When interpreting changes in these measurements, remember that ratios increase either because the numerator increases and/or because the denominator decreases. For example, when ducks given salt water for 2 or 4 days are compared (see Table B.1), over half the difference in salt gland weight (when expressed per unit body weight) can be accounted for by changes in body weight. By contrast (see Table B.2, columns 1 and 3), the increases in salt gland weight in animals that had not been operated upon appear to be underestimated slightly because body weight rose from 4.72 to 5.06 kg, an increase of about 7 per cent.

If changes in body weight are marked, as might occur during periods of rapid growth or when, for some reason, appetite is lost or animals are fed a special diet, then considerable difficulties in interpretation might arise. In these circumstances, weight changes relative to 'control' animals (that is, animals with sham operations, unchanged diets, etc.) would act as a valuable interpretive guide. In the present experiments upon ducks only one control group of animals is used (allowed fresh water and killed on day 0, or given fresh water for 4 days — see Figure B.1 and Table B.1) and so such comparisons cannot be made. A better protocol for experiment 1 would have involved separate control groups at days 0, 1, 2, 7 and 14 so that the effects of changes in weight due to (say) growth and development could have been taken into account.

B.2 (b). Hyperplasia would be manifest as an increase in the DNA content of the salt glands and hypertrophy as increases in RNA and protein contents. (Note that an increase in weight *alone* might be produced by oedema or cellular swelling due to pathological changes.) If the gland is prepared by homogenizing it and then using aliquots for analysis, it is straightforward to calculate the DNA, RNA and protein content of the whole gland provided that the volumes of each aliquot and the total homogenate are known. That is, this method of presentation of results enables an estimate to be made of the presence of hyperplasia and hypertrophy (correcting the results for changes in body weight if necessary).

Another means by which the amount of substance in a gland or tissue can be described is 'weight per mg of tissue'. This is a useful measure when the amount of tissue is fairly constant (for example, the amount of neurotransmitter stored in a nucleus in the brain or of hormone stored in an endocrine gland), but it is a confusing measure if changes in tissue weight are substantial — or even a major item of investigation. Thus, in a gland showing hypertrophy as well as hyperplasia, whether the DNA content (per mg of tissue) rises or falls depends upon the *relative* increases of cell numbers and cell size. If cells become more numerous but smaller, then 'DNA per mg tissue' would rise; if cells grew also, then 'DNA per mg tissue' might fall.

B.3 (a). There is a marked increased specific activity of DNA when animals have been drinking salt water for 2 days (experiment 1). This indicates increased DNA formation during the previous 3 h and implies cell division (or at the least preparation for it) — hyperplasia. Total DNA is significantly raised by day 2 confirming that hyperplasia has occurred. The further rise in total DNA compared with day 2 suggests that further hyperplasia has taken place in the interval 2–14 days. However, the statistical comparison (with respect to day 0) is of no use here; what is required is a comparison of day 2 and day 14, one that was not made. The increase in total DNA by day 14 (per 100 g body weight) is about 40 per cent of the value on day 0. However, also expressed relative to body weight, the increase in gland weight is about 70 per cent over the same period of time; this increase is about double that of the DNA and suggests that, in addition, cells have almost doubled in size — hypertrophy. However, the gain in weight might be the result of water influx into cells and represent rather some transient or pathological state; an increase in cell protein is a better guide to hypertrophy than weight. Another means of assessing hypertrophy would be to compare the RNA/DNA ratio of the tissue. Note that the ordinate scales of the graphs are such that they give the misleading impression that gland weight and total DNA changes were very similar.

B.3 (b). The data from the figure indicated the following:

(1) There is a significant rise in total DNA over control values by day 2 and the mean value is further raised by day 7 and again by day 14. However, the appropriate statistical tests have not been performed so that the statistical significance (or otherwise) of such changes is not known. It is worth noting that visual inspection of the means and their standard error certainly suggests that values on days 2 and 14 are significantly different.

(2) Only on day 2 is there a significant increase in the rate of DNA synthesis as assessed by the specific activity measurements.

The protocol for the experiments is such that the results cannot be used to describe the time-course of DNA synthesis in any detail. Consideration of the incorporation of thymidine shows that DNA was being synthesized on day 2 but synthesis could have begun as early as just after day 1 and continued up to, but not including, day 7. The 'total DNA' results imply that some DNA had been made between days 1 and 2; the time-course of further manufacture cannot be assessed with any degree of confidence due to the absence of suitable statistical tests. The value of these would be to establish if further DNA synthesis had occurred at any time between days 2 and 7 and between days 7 and 14.

To assess the time-course in more detail would require further groups of animals. What these results do show clearly is that, if only a few groups of animals are to be used, then the times at which they are studied must be chosen most carefully.

B.3 (c). One must assume that *in vitro* results from tissue slices can be compared with *in vivo* results from whole glands, if estimates of the growth and cell division of the gland at additional times are required. The data from Table B.1 indicate that DNA synthesis is occurring on day 4 as well as on day 2. The latter time confirms the results of Figure B.1, and the former would offer an explanation of the further rise in total DNA seen in this figure after day 2.

However, there is evidence that the *in vitro* and *in vivo* results are not the same. The incorporation of thymidine was measured after 2 days salt water in both protocols (see Figure B.1 and Table B.1). Two differences exist. First, the 'baseline' values for incorporation of thymidine differ: in the *in vivo* study, they are not appreciably above zero, unlike in the *in vitro* study. Second, the rate of incorporation on day 2 in the *in vitro* experiment is about double that found *in vivo*. There is the further point that the weight of the glands studied *in vitro* has more than doubled by day 2, but it rose by only about 50 per cent in the *in vivo* protocol. This difference seems to arise not only because the *in vitro* control values — animals not exposed to salt water — are lower but also because the weights on day 2 are higher. Thus some doubt must exist as to the validity of the comparisons that have been made by us.

B.4. In all cases, one is investigating the possibility of rejecting the null hypothesis on the grounds that the difference between the groups is unlikely to have arisen by chance — that is, it is a real difference. Some of the implications of finding a significant difference will be mentioned when each hypothesis is considered below; in other cases, further comment is made in relation to question B.5. The null hypotheses are:

A. That salt glands from the left and right sides do not differ in weight in birds drinking fresh water, that is, under control conditions. In other experiments, it is always the left salt gland that is removed so the effects upon these operated animals of drinking water can be investigated in the right salt gland only. Presumably, with intact animals also, the right salt gland is used. Had the two glands differed in weight, this would have greatly increased the complexity of the problem and would have required either separate studies for the left and right glands or the random choice of side in any animal.

B. That the weight of the remaining salt gland of birds given fresh water does not change after unilateral salt gland removal.

C. That, in intact birds, the type of drinking water does not influence the salt gland weight.

D. That, in intact birds given salt water, the weight of the two glands does not differ. This compares the *response* of both glands to salt water (compare with hypothesis A), and the comments made there apply here also.

E. That, in operated birds, the weight of the remaining salt gland does not depend upon the type of drinking water. This is to be compared with the result in intact animals (see hypothesis C). Note that comparisons C and E give the same result, namely a rejection of the null hypothesis. They do *not* compare the size of any changes (approximate doubling in weight versus tripling). Such a comparison of the changes in gland weight in intact or operated geese given salt water or fresh water could be performed by using matched sets of four animals. One animal from each set would be operated upon and given salt water and another left intact but given salt water. The third animal would be operated upon and given fresh water and the fourth left intact, but given fresh water. The ratio of salt gland weights for the animals on salt water to that for the animals on fresh water could be calculated separately for the intact and the operated pairs in each set. This calculation would be performed for all sets of four animals to see if the average ratios differed significantly between intact and operated animals. Alternatively, a two-factor analysis of variance (salt water versus fresh water and operated versus intact) could be performed upon results from groups of animals treated as above.

F. That, in birds drinking salt water, the weight of the right salt gland does not depend upon the presence of the left salt gland. This hypothesis can be compared with B above.

A comparison between intact birds given fresh water and operated birds given salt water is not useful, since any difference could not be ascribed unequivocally to the operation or to the fluid drunk (or to both, as results of testing hypotheses F, E and C imply).

B.5. If a portion of tissue is removed then the remainder often shows hypertrophy and hyperplasia. Two main theories to account for this have been advanced:

(1) the remaining tissue has more work to perform and this is the stimulus for growth — this can be called the 'functional demand' theory;
(2) the cells grow independent of work done because there is less tissue remaining to secrete a chemical substance that inhibits cell division and growth; these growth-inhibiting hormones are some-times called 'chalones'.

If 'functional demand' is important, then there should be an increase in gland size when birds on a salt water diet are compared with birds of the same status (intact or operated) on fresh water diet. This is being tested in the experiments upon ducks (see Figure B.1 and Table B.1) and in hypotheses C and E in the experiments on geese (see Table B.2). The results imply that 'functional demand' does produce hypertrophy and hyperplasia. If, on the other hand, the removal of salt gland tissue is important then, for the same imposed load, there should be similar changes. This is looked for in the hypotheses B and F (see Table B.2). The results here imply that a loss of tissue is a stimulus to hyperplasia and hypertrophy *only* when salt water rather than fresh water is being drunk.

   A difficulty of interpretation exists when these last results are considered. If only one salt gland exists (as after surgery) then any load imposed upon the remaining gland by drinking will be doubled. (Of course, if there is no load, as when drinking fresh water, then this consideration does not apply.) Thus, the present results suggest an effect of tissue mass, at least when salt water is being drunk, but they might reflect instead the effects of a larger or longer-lasting load in operated animals when compared with the case when both salt glands are present. Further experiments using the same loads *as delivered to the salt gland(s)* are required to elucidate this problem.

B.6 (a). For red blood cells, possible markers are radioactive chromium and the surface agglutinins (indeed, the latter marker is used in the Ashby method for estimating the life span of an erythrocyte). For other cells, one problem is that of labelling them specifically in the first place (this is easy when blood is concerned, for it can be removed by venepuncture). The recent development of monoclonal antibody techniques has allowed the labelling of specific types of cells. If such antibodies attach irreversibly and are not recycled following the death of the cell, then they might be useful. In all cases, the rate at which the cells die can be estimated by measuring the rate of loss of the marker. If the marker is radioactive, then the loss might be estimated by non-invasive means not requiring the sacrifice of the animals. To follow the decline in radioactivity after a pulse of labelled thymidine would be useful only if, after DNA breakdown, it was not reincorporated into new DNA.

B.6 (b). Hyperplasia should be detectable by counting the number of cells in a section of salt gland undergoing mitosis. This requires the chromosomes to be arrested at the metaphase stage by adding colchicine and the section to be stained ready for microscopic inspection. As with red cell counts, a standardized procedure for counting becomes necessary. One possible scheme is:

   (1) choose X different fields of view (a microscope-slide with a fine grid can be useful);
   (2) count the number of mitotic spindles in each field of view.

It might be of interest to know that the results should follow a Poisson distribution.
   Hypertrophy requires an estimate of cell volume. This has to be calculated from the profiles seen in histological sections. Clearly, the

planes of section through different cells will vary widely but, with enough cells sampled, these variations will become randomly distributed so that different samples can be compared statistically. The volume of each cell might be calculated from one or more measurements made of each cell as it appears in the section. This will require some mathematical model to describe the relationship between these measurements and the cell volume. For example, it might be

$$\text{volume} = (l_1 \times l_2)^{3/2}$$

where l_1, l_2 are two linear measurements of the cell made at right angles to each other, or

$$\text{volume} = \text{area}^{3/2}.$$

However, there are many potential problems. For example,

(1) The cells are neither spheres nor cubes, nor any regular shape; therefore any equation relating volume to another measurement will be an approximation.
(2) Whatever shape the cell is, the histological section would be unlikely to pass through a particular plane — for example, through a diameter of a sphere. Therefore, the dimensions of the cell section have an unknown relationship to the cell as a whole.
(3) There might be a tendency for large cells (or large sections of cells) to be measured rather than small cells or sections.

As a result, it might be most appropriate to measure the area of each cell profile and to use a statistical test suited to non-parametric data. Further, measuring the areas of cell sections makes few assumptions about their relationship to cell volume except that it is positive. The cumulative frequency distributions of areas for control and experimental tissue can then be compared by the Kolmogorov–Smirnov test to see if they differ significantly.

Abstract C

The effect of stimulation of the nucleus ambiguus upon the plasma concentration of insulin

Introduction

Several studies have shown that stimulation of the peripheral end of the vagal nerve will increase the concentration of insulin in the blood. However, the origin of the vagal efferent fibres which mediate this effect is unknown. Histological studies using horseradish peroxidase which is transported along axons towards the cell bodies suggest that neurones in both the nucleus ambiguus (N.AMB) and the dorsal motor nucleus of the vagus give rise to axons which innervate the pancreas.

This study was designed to investigate the effect of electrical stimulation of the N.AMB on the concentration of circulating insulin. In addition, some of the pharmacological properties of this system were investigated.

Methods

Experiments were carried out on male rats which, until the day preceding the experiment, had been given free access to food and water. Each animal was then housed separately and was not allowed access to food for 16 h before being anaesthetized.

After induction of anaesthesia with pentobarbitone ($30 \, mg \cdot kg^{-1}$ intraperitoneally), a tracheostomy was performed and a catheter was inserted into the right jugular vein. The animal was maintained under anaesthesia and body temperature was controlled throughout the experiment. Heart rate was also monitored throughout.

The dorsal surface of the brainstem was exposed and was kept moist with Ringer's solution. A bipolar electrode was lowered into position at least 30 min before stimulation. At the end of each experiment, a small electrolytic lesion was made via the stimulating electrode and the brain was removed. The site of stimulation could be verified by histological techniques.

Except for the control group (see later) each experiment consisted of two trials with the same electrode position being used for each. During each trial, blood samples were collected at -4, -2, 0, 1, 2, 3, 5 and 10 min (where times before stimulation are denoted by a negative value). Electrical stimulation started at time 0 min and was stopped immediately after the sample at 3 min. After the last sample for trial 1 (at time 10 min), a 30 min waiting period followed before blood sampling was begun for the second trial (-4 min). The detailed protocol was varied between four main groups:

(1) a control group with only one period of electrical stimulation and without any pharmacological agents;
(2) electrical stimulation (trial 1) followed by identical stimulation in the presence of the α-adrenergic antagonist, phentolamine (trial 2);
(3) electrical stimulation in the presence of phentolamine before (trial 1) and after (trial 2) bilateral cervical vagotomy;
(4) electrical stimulation in the presence of phentolamine before (trial 1) and after (trial 2) treatment with atropine.

Blood samples were centrifuged and the concentrations of glucose and of immunoreactive insulin (IRI) in plasma were determined.

A number of brainstem locations were electrically stimulated. The position of the electrode was then categorized into one of two anatomical groupings for the purpose of analysis. The first group consisted of those electrode positions within 500 μm of the nucleus ambiguus (N.AMB). The second group was referred to as the control group (CONT.) and consisted of those electrode positions further than 500 μm from the nucleus ambiguus but not impinging on other known vagal nuclei.

The IRI response for each animal was categorized in terms of the percentage change during the 3 min of brainstem stimulation (0–3 min) compared with the initial control values (from -4 to 0 min). These categories were: < 30 per cent IRI increase; 30–90 per cent increase and > 90 per cent increase.

Only results from stimulation parameters of strength 50 μA, frequency 30 Hz and duration 0.2 s are presented. At greater stimulus strengths, severe depression of heart rate and respiratory rhythm were seen when stimulation was in the region of the nucleus ambiguus; such stimulation also had inconsistent effects on the plasma concentration of IRI.

Values are expressed as means $\pm$ SE followed by the number of animals in parentheses unless otherwise stated.

Results

Controls (group a)
In no case did electrical stimulation from an electrode within the N.AMB group give an IRI increase of < 30 per cent and 25 of 30 positions included in this group yielded a > 90 per cent increase in the concentration of IRI. Conversely, of those electrode positions classified as in the CONT. group, only two of the 40 caused a > 90 per cent change in IRI concentration during the 3 min of stimulation, and six caused a 30–90 per cent increase.

Before stimulation, values for plasma concentrations of IRI and glucose did not differ significantly between N.AMB and CONT groups. The changes in plasma concentrations of IRI and glucose from the initial control values (of 2.3 ± 0.1 (70) ng·ml^{-1} and 127.4 ± 1.8 (70) mg·dl^{-1} respectively) are shown in Figure C.1.

The rise in IRI following stimulation of the N.AMB group was accompanied by a small but insignificant fall in heart rate, − 2.1 ± 7.2 (30) beats·min^{-1}, mean ± SD (number of results), and there was no noticeable change in respiratory frequency or rhythm.

Treatment with phentolamine (group b)
Phentolamine raised the initial control plasma concentration of IRI from 2.1 ± 0.2 (24) ng·ml^{-1} to 8.6 ± 1.1 (24) ng·ml^{-1} but had no significant effect on the plasma concentration of glucose (129.3 ± 4.6 (24) *vs.* 123.3 ± 2.7 (24) mg·dl^{-1}). Heart rate was not altered by phentolamine.

The effect of brainstem stimulation on plasma IRI and glucose in phentolamine-treated animals is shown in Figure C.2.

Vagotomy and treatment with atropine (groups c and d)
Neither vagotomy nor treatment with atropine prior to trial 2 altered the concentrations of IRI or glucose when compared with initial control values seen after phentolamine treatment alone (trial 1). The results showed that the rise in IRI following electrical stimulation of the N.AMB group could be greatly diminished following vagotomy or treatment with atropine. With both experimental procedures, the concentration of glucose was not significantly changed when compared to values observed during trial 1.

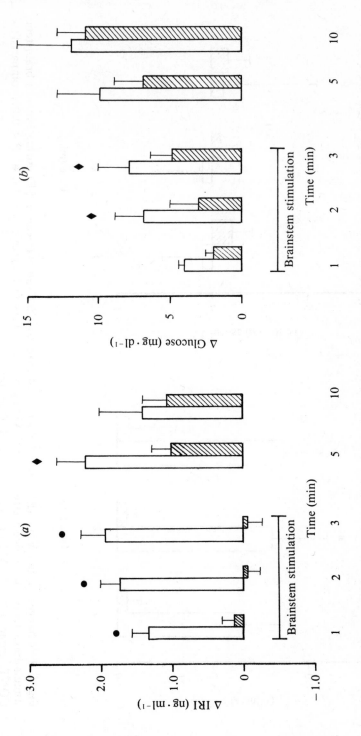

Fig. C.1　Changes (Δ) in plasma IRI (a) and glucose (b) from baseline values (mean ± SE) during each of three minutes brainstem stimulation and at 5 and 10 min (post-stimulation); N.AMB group, $n = 30$; CONT. group, $n = 40$. ♦, $P < 0.05$; ●, $P < 0.01$, compared to CONT. group. □ = N.AMB group; ▨ = CONT. group.

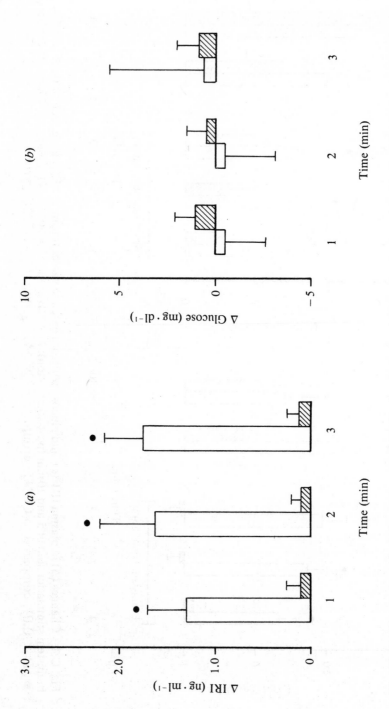

Fig. C.2 Changes (Δ) in plasma IRI (*a*) and glucose (*b*) from baseline values (mean ± SE) during 3 min of brainstem stimulation after phentolamine pretreatment; N.AMB group, $n = 11$; CONT. group, $n = 13$. ●, $P < 0.01$, compared to CONT. group. ☐ = N.AMB group; ▨ = CONT. group.

Discussion

The increase in the concentration of IRI as a result of electrical stimulation of the N.AMB group was probably due to direct activation of vagal motoneurones, but stimulation of fibres coming from other vagal motor nuclei (such as the dorsal vagal nucleus) and passing through the region of the nucleus ambiguus cannot be excluded.

On the basis of these experiments alone, it is not possible to conclude that the rise in the concentration of IRI produced by stimulation of the nucleus ambiguus is a direct effect mediated by the vagal innervation of the pancreas. For example, an alternative explanation would be that the stimulation could have activated vagal motoneurones controlling the release of gastrointestinal hormones which could then cause the release of insulin.

Questions

C.1 How would you assess whether the preparation used in the present study was physiologically normal?

C.2 How adequate do you consider the protocol to have been in these experiments?

C.3 Have you any comments on the technique used to measure the concentration of insulin in the blood?

C.4 Have you any comments upon:
(a) the division of electrode positions into N.AMB and CONT. groups;
(b) the stimulus parameters chosen for the present study?

C.5 Outline the statistical tests you would use and the data you would need to investigate:
(a) if the IRI changes after N.AMB stimulation of controls (group a) depended upon the area of the brain being stimulated;
(b) if stimulation of the N.AMB produced a significant change in heart rate in this group.

C.6 Consider how the initial suggestions made by the researchers in the introduction might be substantiated or modified by the results obtained in these experiments.

C.7 Briefly describe how you would investigate the comments and suggestions given in the discussion.

Suggested answers

C.1. Variables such as blood gases, blood pH, body temperature, arterial blood pressure and plasma osmolarity should be controlled, of course. Evidence is presented that some of these variables (temperature, heart rate and respiratory rhythm) were monitored and that the experimental protocol did not cause them to change. Further variables, more directly related to this particular experiment, should be measured, if possible. An example (which was carried out in these experiments) is ensuring that control concentrations of plasma glucose and insulin are within the normal range or, at the least, any changes are known. By contrast, the environment of the brain — in particular the area around the vagal nuclei — was not monitored. This might have involved checks on c.s.f. composition, especially pH, or even brain electrical activity (the e.e.g). Although it probably is not feasible to do this in every animal, it could have been done in some initial pilot studies.

In addition to establishing that control values were within the normal range, it would have been desirable in some preliminary experiments or sham-operated controls to ensure that the anaesthetic, surgery and general protocol did not alter β-cell or vagal function. The former can be assessed by a glucose tolerance test. A method for testing the latter is not so obvious, but vagally mediated changes in gastrointestinal motility or secretion might be useful.

C.2. Many aspects of the protocol are commendable.

Thus there are control blood samples (-4, -2, 0 min) and frequent blood samples during (1, 2, 3 min) and after (5, 10 min) electrical stimulation. However, no mention is made of the size of the sample removed and therefore it is not known whether or not this represents a significant haemorrhage to the animals. The electrical stimulation was standardized with respect to the characteristics of the pulse, its frequency and duration and it was delivered to the same area of the brainstem in the two trials. Further, there was not only a 'recovery period' after electrode penetration and before stimulation but also between successive bouts of stimulation. However, there is no evidence given to show that variables had returned to control values before the next stimulation period was begun.

Notice that, after phentolamine pretreatment, the animals now have raised concentrations of IRI but unchanged values for plasma glucose concentration and heart rate. These (sometimes changed) values become the new controls for the second stage of the experiments; it is against these that the effects of brainstem stimulation are now to be judged. Similarly, 'abnormal' baselines exist when vagotomy and atropine treatment are used with groups c and d.

The general division of the protocol into two trials, with the first acting as a control for the second, is also commendable and allows statistical tests suitable for paired data to be used. Note that, ideally, a better arrangement would have been either to decide randomly between the order of trials 1 and 2 or to have repeated trial 1 after trial 2. This second suggestion can result in long experiments and so increases the likelihood of deterioration of the animal preparation. The first suggestion can pose difficulties after the administration of drugs (the 'carryover' effect) and is not possible after a surgical vagotomy; for this reason, temporary vagotomy (by cooling or by the application of local anaesthetic) might be preferred.

Note that group a differed from groups b–d by virtue of undergoing only one trial. It would have been better for this group also to have undergone two (identical) trials. This would have established if a second period of stimulation produced results identical to the first. Unless this is the case then the interpretation of results from trial 2 in groups b–d is suspect.

C.3. Note that the interpretation of the results depends critically upon the validity of the immunoassay method for insulin. In such cases, some indication of the validity of the method is required (it is not included in the present abstract). It enables the reader to ascertain how reproducible is the analysis (by performing it a number of times upon the same sample or standard) and how specific (ensuring it picks up *all* of the insulin but *none* of the other proteins). There are similar problems when the glucose assay is considered. Again, evidence for its reproducibility and specificity (only D-glucose, or all hexoses, or only those with a particular grouping at the C_2-atom?) would be required (see Chapter 5).

C.4. The division into N.AMB and CONT. groups on the basis of the position of the stimulating electrode necessarily requires some assumptions to be made. First, it assumes that the nucleus is a clearly defined structure rather than an anatomically diffuse grouping of cells, albeit of specific function. Second, it assumes that all cells within a certain radius of the electrode (? $10 \mu m$) — but only these cells — will be stimulated. With extracellular stimulation the currents involved will spread away from the electrode position by an amount that is presumably related to current strength but in a complex (unknown) way. Moreover, neurones with large cell bodies will be stimulated in preference to (that is, at a more distant site from the stimulating electrode than) those with smaller cell bodies.

If the distance from the nucleus ambiguus is set at too low a value, then occasions will arise when cells that function as part of this nucleus are categorized as CONT.; the opposite problem applies when the distance is set at too large a value, of course.

The stimulus parameters also require careful consideration. Sometimes the parameters have been standardized by previous researchers, but often the choice is determined by the present experimenters. Decisions have to be taken regarding the shape of each pulse and its strength, the frequency and duration of bursts of impulses, and the interval between such trains.

In the present experiments, the parameters were chosen in part because stronger stimuli gave evidence of having spread too far to areas of the brainstem with respiratory and cardiovascular functions and of producing 'inconsistent' effects upon plasma concentrations of IRI. In favour of this decision is the fact that such effects would be difficult to reconcile with the idea of a 'specific' stimulus, as was required in these experiments. By contrast, the results might indicate that other mechanisms (more distant? of higher threshold?) that inhibit insulin secretion might exist. These mechanisms might, in practice, be an *integral part* of the response and to design the experiment so that they were not invoked might limit the physiological relevance of any results that are obtained. The choice of frequency and duration of stimulation should be determined very considerably by the values found normally and this requires recordings to be made *in situ*. Until such data are known, the values chosen might not relate to any reasonable physiological circumstance, again limiting the usefulness of the study.

C.5 (a). The results are presented in a nominal form suitable for χ^2 analysis. It should be realised that the abstract indicates that IRI was measured as $ng \cdot ml^{-1}$, that is, on an interval scale. Therefore, the results would appear to be amenable to statistical testing by a parametric test (Student's t test) or, if not normally distributed, by an ordinal test (Mann–Whitney U test). Thus N.AMB and CONT. groups could have been compared in these ways. In connection with this, note that the calculations of mean $\pm$ SE displayed in Figures C.1 and C.2 assume that the results have achieved interval status. However, for some unknown reason, the results have been presented in a format appropriate to a nominal statistical test only. The advantage (if any) of this is unclear. However, the 3×2 contingency table is shown below:

IRI change	N.AMB	CONT.
< 30 per cent	0 (13.7)	32 (18.3)
30−90 per cent	5 (4.7)	6 (6.3)
> 90 per cent	25 (11.6)	2 (15.4)

Expected values, assuming a null hypothesis (that no difference between N.AMB and CONT. groups existed) are shown in brackets. Note that less than 20 per cent of the expected results are less than 5, so that there is no need to combine categories.

The null hypothesis can be rejected at the 1 per cent level. This division of IRI changes into the categories used is arbitrary, both with respect to the number of divisions and their borderlines. In the absence of any such guidelines from the published literature, one might categorize the data such that, considering all the results from both groups ($n = 70$), an approximately equal distribution between the categories would be achieved. However, as the table shows, few results from either group were found in the 30−90 per cent category. It therefore serves little purpose; either it should be widened or removed altogether. As the number of categories of either variable is increased then more precise hypotheses can be tested. For instance, one might consider if electrode positions 300−500 μm from the nucleus ambiguus had the same effect as those 100−300 μm distant. However, the number of results in each category decreases and this can lead to difficulties when statistical tests are applied and interpretive problems might multiply alarmingly with an increase in categories.

C.5 (b). The null hypothesis is that there is no significant change in heart rate, that is, the mean change was not significantly different from zero. To test this requires the standard error of the sample to be calculated:

$$\begin{aligned} SE &= SD/\sqrt{n} \\ &= 7.2/\sqrt{30}. \\ &= 1.3 \end{aligned}$$

Then the value of t is calculated from:

$$t_{29} = \frac{2.1}{SE} = 1.59$$

where 2.1 is the sample mean and there are 29 degrees of freedom. Note that t rather than z is calculated since the variance of the population had to be estimated from the sample. The null hypothesis cannot be rejected at the 5 per cent level. Therefore we conclude that no significant change in heart rate was produced by stimulation of the N.AMB group. Even so, it is instructive to estimate the spread of heart rate changes. Assuming them to be normally distributed, approximately two-thirds would lie in the range -9 to $+5$ beats $\cdot$ min^{-1} (mean $\pm$ 1 $\times$ SD) and 95 per cent in the range -16 to $+12$ beats $\cdot$ min^{-1} (mean $\pm$ 2 $\times$ SD).

C.6. The initial suggestion was that there were nerve cells in the N.AMB which, when stimulated, would cause insulin release. Presumably, the released insulin would, in due course, lower the concentation of glucose in the blood. The controls (group a) gave partial support to this view (see Figure C.1) since N.AMB, but not CONT., stimulation raised the concentration of IRI. However, there were some problems. There was a rise in plasma glucose concentration after stimulation of both areas but *particularly* after N.AMB stimulation. Further, concentrations of IRI were raised at 5 and 10 min with CONT. stimulation; even though the latter could be explained as a response to the raised plasma glucose, there was a *fall* in IRI levels between 5 and 10 min with N.AMB stimulation, even though plasma glucose concentration continued to rise. (Indeed, at 10 min IRI concentrations in N.AMB and CONT. groups are not significantly different.) One model which could account for many of these results is shown in Figure C.3.

Since phentolamine treatment (group b, trial l) raised the concentrations of IRI (incidentally, by about three times that produced by N.AMB stimulation in group a) but not that of glucose, this suggests that there is a tonic inhibition (via α-receptors) of insulin release but that there is no such tonic effect upon plasma glucose concentration. Also (trial 2), since plasma glucose concentration does not rise in this group with stimulation, it implies that the rises in group a were mediated

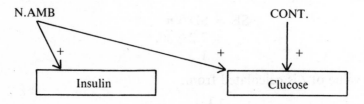

Fig. C.3   One model to explain the date presented in Figure C.1.

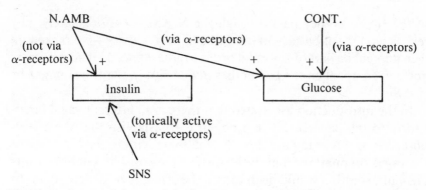

Fig. C.4   One model to explain the data presented in Figures C.1 and C.2.

via α-receptors. Finally, the IRI response was unchanged in size (compare Figures C.1 and C.2 implying that this was not mediated by α-receptors). The model can thus be changed to that shown in Figure C.4.

The experiments upon groups c and d are complementary. Since vagotomy or treatment with atropine did not alter the concentrations of IRI or glucose prior to trial 2, this indicates that any vagally mediated, cholinergic pathways that affect either IRI or glucose release are not tonically active. The results of trial 2 indicate that these pathways are not the only ones (assuming that vagotomy or cholingeric block were complete) which mediate the effects of stimulation upon IRI and that they do not affect glucose concentrations. Therefore, further amendments to the diagram can be made.

Clearly, other interpretations are possible: the above models merely illustrate the way in which hypotheses can be modified (or 'refined') as data are gathered. Their usefulness is in enabling a 'synthesis' of results to be made and in suggesting further experimental tests.

C.7. It is required to show that the electrodes have not stimulated nerve fibres merely passing through the N.AMB just as (see question C.4) it was important that the stimulus current did not spread too far. This is an exceedingly difficult problem to deal with; it is an example of the difficulties of interpretation that can arise when the techniques available are limited. Thus, we cannot yet reliably stimulate cells rather than fibres by electrical means; this problem will remain unless or until we can. However, the iontophoretic application of neurotransmitters might be a useful approach. If fibres originating from the dorsal vagal nucleus are suspected, then recordings made from this nucleus might establish whether or not cells have been activated antidromically after stimulation of the fibres as they pass through the N.AMB. Unfortunately, any results would not be easy to interpret since if activity were picked up then it might be part of a normal orthodromic input and if activity were not picked up, then the cell bodies giving rise to the fibres might be elsewhere.

In the introduction it is stated that there is an efferent vagal innervation to the pancreas. The problem is to establish that the effects observed in this study are directly due to activity in the branches of the vagus originating from the N.AMB. One possible, but technically difficult, solution would be to record the efferent vagal activity in the efferent vagus close to the pancreas during N.AMB stimulation (it would be expected to be the same as the stimuli delivered at the N.AMB); then one could stimulate the pancreas from this peripheral site at the same frequency to see if the same results were obtained. If they were, then other factors, such as blood-borne agents, are not required to account for the effects of N.AMB stimulation. To repeat this protocol, but when stimulating the dorsal motor nucleus of the vagus, would be instructive. The more general possibility that there was a blood-borne agent (from the gut?) acting as a secretagogue could be tested by radioimmunoassay of the blood for possible factors, such as glucagon, during the trials. For such a factor to be an acceptable mediator of the observed responses, the time-course of its concentration in the blood passing to the pancreas would have to be appropriate for the size of effect that was produced in the present experiments. Such an approach depends on there being an appropriate assay available for each substance. In practice, this would rarely be done since some of the possible mediators may be unknown. However, an experimenter may choose to assay one or more particular substances if he has reason to suspect their involvement as mediators from previous studies. It is worth noting that the effects of electrical stimulation, though rapid in onset, seem to outlast the stimulus (see Figure C.1); this accords with the concept of a slower decay of a humoral, rather than a neural, agent. However, it might, instead or in addition, reflect the time-course of destruction of IRI or the half-life of its action on the cell rather than its release. In addition, a vagally denervated pancreas would respond to a blood-borne agent, but not to a direct vagal influence. Such a clear-cut distinction would not apply if both mechanisms were normally involved, for example, if the neural input to the pancreas determined its sensitivity to humoral agents (compare the interaction of the vagus, gastrin and mechanical factors at the site of secretion of gastric juice).

The response of medullary neurones to mechanical loads imposed during inspiration

Introduction

When inspiration is inhibited, there is immediately a reflex increase in the activity of the external intercostal muscles and diaphragm. The response of the external intercostal muscles includes

(1) an increased rate of motor unit activity;
(2) an earlier excitation of motor units;
(3) a more prolonged discharge from the motor units; and
(4) the recruitment of new motor units.

Recently, the afferent pathways mediating these changes have been described. Changes in the discharge of vagal afferent fibres contribute to the increased rate of motor unit activity and the recruitment of new motor units, and are entirely responsible for the increase in the duration of discharge from the motor units. Afferent information reaching the central nervous system via the thoracic dorsal roots contributes to the increased rate of motor unit activity and the recruitment of new motor units and accounts fully for the earlier excitation of motor units. The responses of the diaphragm are an increase in the rate of motor unit activity and a prolonging of the duration of discharge, and both these effects depend upon afferent information reaching the central nervous system via the vagal nerves.

It is not known how this afferent information is processed in the central nervous system, particularly to what extent it influences the discharge of medullary inspiratory neurones. The present investigation was carried out in order to investigate this problem by recording from inspiratory neurones in the medulla. Since there is some evidence that the reflex response to a fall in total compliance differs from that to an increase in airway resistance, both kinds of inspiratory load have been tested.

Method

Experiments were performed on cats anaesthetized with α-chloralose (80 mg · kg^{-1}). Tracheostomy was performed routinely on all animals and in some cats vagotomy was also performed. Rectal temperature was monitored and maintained within normal limits. All experimental data presented are from animals whose blood pressure, temperature and blood gases were normal. Air flow in the trachea was recorded.

Medullary inspiratory neurones were selected by the following criteria: the units fired during inspiration showed a pattern of firing with a time-course similar to that of airflow and were inhibited by lung inflation. Electrical activity from these medullary neurones was recorded during inspiration.

An elastic load (mimicking a fall in lung compliance) or resistive load (mimicking an increase in airway resistance) was connected to the tracheal cannula. The resistive load was produced by making the animals breathe through porous metal discs and the elastic load was provided by making the animals breathe air from a large glass flask.

The following protocol applied to all experiments. A control recording was taken from a single medullary inspiratory neurone and then the cat was subjected to the inspiratory load. The load was always applied for one inspiration only, starting at the beginning of that inspiration. The load was added six times. Between loads, the cat was allowed to take ten breaths, during which time breathing returned to the previous control pattern.

The effect on neuronal activity was determined by comparing the average (from the six trials) of the total amount and duration of the neuronal activity during the loaded inspiration with that during the three previous unloaded inspirations.

Results

A summary of all the results is presented in Table D.1. Three neurones began firing late in the course of control inspirations and none of these showed a change in the time of onset of firing during a loaded inspiration.

An example of a response during resistive loading is shown in Figure D.1. The animal displayed the classical response to vagotomy — that is, increased tidal volume, decreased frequency of respiration and an increased breath-to-breath variability in these variables. After vagotomy,

Table D.1 Effects of resistive and elastic loading upon rate and duration of firing of medullary neurones

Preparation	Load	Number of units	Rate ↑	Rate ↓	Rate ↔	Duration ↑	Duration ↔
Vagi intact	Resistive	39	19	1	19	36	3
	Elastic	22	10	6	6	21	1
Vagotomy	Resistive	18	0	3	15	1	17
	Elastic	15	1	4	10	1	14

↑ , significant increase, $P < 0.05$
↓ , significant decrease, $P < 0.05$
↔, no significant change

there was no difference between the rate and duration of firing in control and loaded breaths. However, in animals in which the vagi were intact, there were statistically significant differences observed in both duration and rate of firing.

Discussion

The results show that activity in medullary inspiratory neurones is altered immediately on mechanical loading of inspiration. The time-course of such a response, and its absence after vagotomy, establishes the dependence of the response on neural inputs and eliminates any contribution from chemical changes.

The predominant effect of resistive loading on medullary inspiratory activity was an increase in both rate and duration of neurone activity. The mechanism by which information carried by vagal afferents facilitates medullary inspiratory activity after a resistive load probably involves pulmonary inflation-type receptors. Thus, addition of a resistive load during inspiration would decrease the rate of lung inflation; since this would represent a reduction in inhibitory information impinging on the inspiratory neurones, a facilitation of their activity would result. The absence of facilitation of medullary inspiratory neurones in the vagotomized animal indicates that the vagi are the only source of sensory information facilitating medullary inspiratory activity during resistive loading.

(a) *Intact animal*

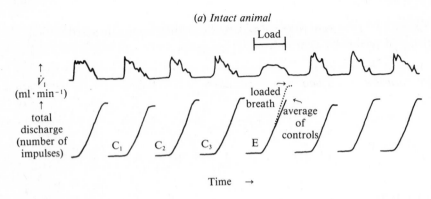

Time →

(b) *After bilateral vagotomy*

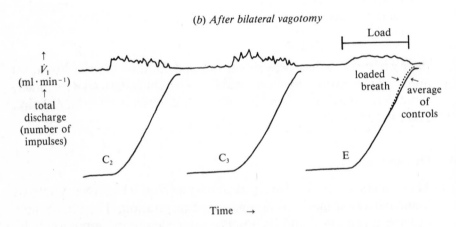

Time →

Fig. D.1 *(a)* the firing responses of one neurone during resistance loading with the vagi intact; and *(b)* of another neurone after bilateral vagotomy. The top trace is inspiratory flow rate and the bottom trace is the discharge of the neurone integrated with respect to time. $C_1 - C_3$ are control breaths, E the loaded breath.

In the animals with intact vagi the responses of medullary inspiratory neurones to elastic loading were an extended firing time and, frequently, an increase in rate of activity. Both changes were eliminated by vagotomy. However, elastic loading also produced a decrease in the activity of a substantial proportion of medullary inspiratory neurones; this inhibition was not altered by vagotomy. Possible sources of such an inhibition are the tendon organs of the diaphragm and external intercostal muscles.

Questions

D.1 *(a)* By means of a diagram, suggest the pathways which are thought to be involved in the four different responses of the external intercostal muscles to inhibition of inspiration.
(b) After consideration of the results section, suggest how this diagram might be altered.

D.2 To what extent is the animal preparation adequate for the present investigation?

D.3 Are the criteria by which inspiratory neurones were selected sufficient for their positive identification?

D.4 Comment on the adequacy, or otherwise, of the present experimental protocol. Does it enable the assumptions described in the first part of the discussion to be made?

D.5 Comment on the results and outline a statistical assessment of them where appropriate.
Consider in particular:
(a) how the data, as presented in Figure D.1, enable changes in neuronal activity to be assessed;
(b) how the results in Table D.1 were obtained and how they can be analysed.

D.6 Comment on the explanation put forward for the difference found between the responses to elastic and resistive loading. What further experiments could be performed?

Answers

D.1 (a). One simple scheme is shown in Figure D.2. A slightly more detailed explanation of events occurring at the motor units might produce a model which describes effects upon:

spike frequency (firing rate)
e.p.s.p. duration (firing length)
e.p.s.p. rise time (start of firing)
low-threshold and high-threshold units (number firing)

Neither the introduction nor these models draw a distinction between the inhibition of inspiration as caused by resistive or compliance loads.

D.1 (b). The results of the present abstract enable a distinction between the effects of raised resistance and decreased complaince to be made in terms of some of the effects upon firing patterns of medullary neurones. They consider the role of vagal afferents only. One possible model is shown in Figure D.3.

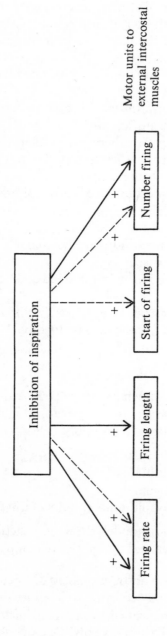

Fig. D.2 Pathways which are thought to be involved in the four different responses of the external intercostal muscles to inhibition of inspiration: ——— mediated by vagi; – – – mediated by thoracic dorsal roots.

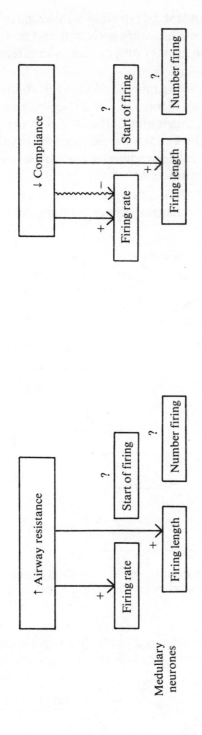

Fig. D.3 Model to show the possible pathways by which inspiratory loads influence medullary neurones: ⌇⌇⌇, not mediated by vagi; ——, mediated by vagi; ?, not assessed.

Note that the present protocol did not allow an assessment to be made of the number of medullary neurones activated and that there were insufficient results (three neurones only) to consider effects upon the onset of firing.

In some way, this part of the model (Figure D.3) should be incorporated into the first one (see Figure D.2), at a stage between inhibition of inspiration and effects upon motor units. However, the question as to the pathway by which the thoracic dorsal roots might influence the motor units — via a spinal reflex mechanism and/or via relaying in the medulla — has not been considered.

A possible further model which concentrates more upon the *difference* between effects produced by changing airway resistance and compliance is shown in Figure D.4.

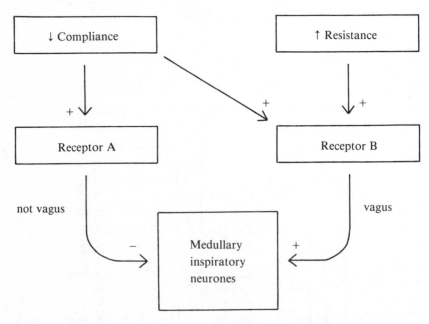

Fig. D.4   Another model to show the possible pathways and receptors involved in mediating the responses of the medullary inspiratory neurones to different inspiratory loads.

As can be seen, all such models require further experimental testing: this is an example of obtaining a set of results that answers one question and poses two more!

D.2. In addition to standard monitoring of blood pressure, temperature, gases, etc., tests are desirable to establish that the brain in general and that respiratory reflexes and the medulla in particular are working satisfactorily. Some possibilities are measurements of the e.e.g., of chemoreceptor reflexes and of the Hering–Breuer reflexes.

Such tests can be both time-consuming and rather elaborate to perform and so they might intrude upon the protocol of normal experiments as performed here. Often, preliminary (or pilot) experiments can be carried out that concentrate upon establishing the health of the preparation as assessed by these tests. Should the outcome of these be satisfactory then the main experimental programme can be undertaken with some assurance that the preparation is physiologically normal. Even so, some test of the normality of respiratory reflexes in particular is desirable during the main experiments. One possibility would be to measure the reflex response to the mechanical load. This response could be assessed as the subatmospheric pressure developed in the tracheal cannula after airway obstruction as well as the change in the electromyogram (e.m.g.) from intercostal muscles. Animals giving rise to values outside a normal range could then be rejected. Further, if the experiment was to last some time, this test could be applied at regular intervals throughout the protocol to test for deterioration in the preparation.

D.3. An early view on respiratory control was that the respiratory rhythm originated in the medulla as a result of the alternating activity between inspiratory and expiratory neurone pools, this activity alternating due to a combination of mutual inhibition, self-re-excitation and fatigue.

More recent work accepts that there are several types of 'inspiratory' and 'expiratory' neurone that vary in their firing patterns. Thus: some fire with a time-course similar to that of inspired volume; others with a time-course in parallel with inspiratory flow rate; others when respiration switches from inspiration to expiration (or vice versa); and others, exerting a braking effect, mainly during the first part of expiration.

The implications of the above are that some neurones involved in the inspiratory process will be excluded by the criteria used here; indeed, only one type of 'inspiratory neurone' will be chosen.

However, it does not necessarily follow that the positive identification of an 'inspiratory neurone' by such restrictive criteria leads to very little chance of error. It might be that the identified neurones are responsible for airway calibre, movements of the nares, etc. instead of stimulating the inspiratory muscles. In addition, there is the practical problem that the result is a recording artefact. The medulla is not a stable recording site and the effects of respiratory movements can cause rhythmic artefacts as the cell moves against the recording electrode.

There are two additional means by which 'inspiratory neurones' could be identified. First, one could stimulate the cell and see if there was either a mechanical or an electrical (e.m.g.) response of the intercostal muscles; however, the effect upon the motor units of stimulating a single medullary cell (rather than a whole group as would normally be the case) might be too small to measure at all easily. Secondly, one could stimulate the fibres in the spinal cord to see if there is antidromic activation of the medullary cells under investigation. Of course, this technique will not be useful if there is a synapse between the medullary cell and the site of stimulation in the spinal cord.

D.4. Many aspects of this protocol are commendable. Thus:

(1) control readings were taken before loading;
(2) the load was always given at the onset of inspiration;
(3) the load was applied for one inspiration only;
(4) the animal was allowed to recover between loads (though evidence that this, in fact, occurred is not given); and
(5) the load was added six times, not once, thus reducing the effects of 'freak' results.

It will be noticed that by this protocol the activity of each neurone is measured under control as well as loaded conditions so that paired statistical techniques can be used when assessing if the load produces a change in firing pattern. However, for technical reasons, the same neurone cannot be recorded from before and after vagotomy and so a paired statistical comparison of the effects produced by loading cannot be made between 'vagi intact' and 'vagotomy' groups. Moreover, in the present experiments, a comparison with the same medullary neurone of resistive and compliance loads was not made even though this presents no technical problems. The loss of information that such lack of 'pairing' produces will be commented upon later.

Note that after vagotomy the 'controls' are now no longer normal animals; note also that, since the breathing was more irregular then the use of more than three control breaths in calculating the average values before loading would have been desirable.

There is another reservation about the protocol that is potentially far more serious. Since the immediate responses to loading are being observed (during the loaded breath itself), it is assumed that they cannot be due to effects via chemoreceptors and that any changes in blood gases have been corrected by the time of the next trial. However, two comments can be made:

(1) No data on arterial blood gases are given.
(2) There will be an *immediate* change in the blood leaving the lungs and the transit time between the lungs and peripheral (especially aortic) chemoreceptors might be less than the time taken for inspiration, at least for the slower breaths after vagotomy.

Information on these points would require the continuous measurement of the composition of the blood passing to the chemoreceptor regions. This would technically be rather difficult since not only are the aortic bodies rather inaccessible, but also the speeds of response of any probes and recording apparatus have to be very fast to be of use.

D.5. The aim of the experiments was to see if any changes in the firing of medullary 'inspiratory' neurones were produced by respiratory loads that had previously been shown to alter the pattern of firing of motor units supplying intercostal muscles. Thus, one might investigate if, after mechanical loading, inspiratory neurones fired earlier, for longer and at a greater rate. (An increase in the number of medullary neurones firing during the loaded inspiration cannot be assessed in the present experiments.)

Only three neurones that normally commenced firing late in inspiration were recorded from. This is too few to establish if neurones in general fired earlier during a loaded inspiration. However, effects upon the duration and rate of firing of inspiratory neurones can be assessed from the results as presented in Figure D.1. (Only the results relating to rate of firing are considered in any detail.) Thus, the total amount of activity is given by the height of the lower trace and the rate of firing by its gradient. The duration of neuronal firing can also be assessed from this trace.

The upper trace gives the rate of airflow ($\dot{V}_I$) at any moment and its area above the baseline (that is, the integral of flow with respect to time) is the amount of air inspired. Notice that, during the loaded breath when vagi are intact, the total neuronal discharge increases at the same time as the tidal volume *decreases*. That is, the increased discharge is associated with an inadequate inspiratory achievement. This accords with the view that the increased neuronal activity is a response to inadequate inspiration (as required by the models; see question D.1). By contrast, had the loaded breath been associated with a larger inspiration, then it would not have been possible to distinguish between the increased neuronal activity producing the larger inspiration or resulting from it.

D.5 (a). A statistical assessment of the results as presented in Figure D.1 can be made with the use of a paired test. Each neurone can be considered individually; thus, the differences between the average rates of neuronal firing during unloaded and loaded breaths can be calculated. The null hypothesis would be that loading produced no change. Such a hypothesis would be rejected if the difference observed would have been expected to have occurred by chance with a probability of less than 0.05. This enables the result from each neurone to be assessed as 'no change', 'rise' or 'fall'. (In the strictest sense, rejection of the null hypothesis only implies 'change' without reference to the direction.)

Since rates are being considered and these might not be normally distributed about the mean value, a paired test for ordinal data — Wilcoxon matched-pairs signed ranks test — might be most appropriate or the results could be logarithmically transformed to 'normalize' them.

The following explains in more detail how the results are to be used. The experimenters proceeded:

control breath 1, C_1
control breath 2, C_2
control breath 3, C_3

Therefore,

average control breath, $\bar{C}$
experimental breath, E.

The response to loading in one test on this neurone was, therefore, $(\bar{C} - E)$. After six tests on the same neurone ($n = 6$), it could be established if the mean value for $(\bar{C} - E)$ differed from zero significantly (in which case the result would be 'rise' or 'fall') or non-significantly ('no change').

A better method exists which makes fuller use of the information obtained and which does not 'downgrade' the results into categories. With this method, we calculate the average value for $(\bar{C} - E)$ from the six tests as above, but then make this the result from *one* neurone. Subsequent neurones could be tested similarly, each producing a value of $(\bar{C} - E)$. Then the average value for $(\bar{C} - E)$ for the group of neurones could be calculated. Note that, if ten neurones are studied, then $n = 10$, even though the total number of experimental (loaded) breaths was 60 and the total number of control (unloaded) breaths was 180. The advantage of testing each neurone more than once is to reduce the effects of breath-to-breath variation. Note also that, with only slight changes to the experimental protocol, analysis of variance techniques could be used and that they would enable the effects of loading, intra-neuronal and interneuronal variation to be investigated.

D.5 (b). As presented in Table D.1, the results have been summarized in a nominal form, the categories being 'increase', 'decrease' and 'no change'. For both rate and duration, separately for resistive and elastic loads, the results can be assessed by the χ^2 test. In practice, many of the expected values will be less than 5, in which cases it will be necessary to combine categories and even to apply Fisher's exact test. When the effect of vagotomy upon the rate of firing in response to a resistive load (expressed as 'increase', 'not increase', that is, combining 'decrease' with 'no change') is considered, the contingency table is shown in Table D.2; $\chi^2 = 13.2$ with 1 degree of freedom. This enables the null hypothesis to be rejected at the 0.1 per cent level. Thus we can conclude that vagotomy does affect the rate of firing of inspiratory neurones in response to a resistive load.

Even though the protocol enables a paired statistical test to be used when the decision is taken as to how to categorize the response of each

Table D.2 Summary of results for chi-squared test derived from Table D.1 (for more details see text)

Firing rate	Intact		Vagotomized	
Increase	19	13	0	6
No increase	20	26	18	12

Expected values (assuming the null hypothesis that vagotomy does not alter the distribution of cells between the categories 'increase' and 'not increase') are underlined.

neurone to loading, there are other occasions when the protocol does not permit the use of such tests. For example, the same neurone was not investigated with both types of loading nor was the same neurone used before and after vagotomy. In the latter (but not the former) example, there are severe technical problems why this could not be done.

This lack of pairing can reduce the information that is derived from the experiments. Thus, the results with elastic loading (in which some neurones increased and others decreased their rates of firing) might have been more informative if they could have been compared with the response of the same neurone to the resistive load. For example, was an 'increase' in firing with an elastic load always associated with an 'increase' in firing with a resistive load; and were 'decreases' in firing with an elastic load associated with 'no change' in firing with a resistive load, etc.? Only a small change in the design of the experiments might have produced a considerable increase in the usefulness of information obtained from them.

D.6. In the discussion, it is suggested that the facilitation of medullary inspiratory neurones following resistive loading is due wholly to vagal inputs originating from pulmonary inflation-type receptors. With elastic loading, there is an additional inhibitory input from tendon organs mediated via a non-vagal pathway (see also Figures D.3 and D.4 in answer D.1(b)). With this latter type of inspiratory load, it is not known if the two inputs impinge upon different medullary inspiratory neurones (thus accounting for the different responses) or if each neurone 'integrates' its afferent input (thus giving a range of responses, including 'no change'). If the 'integrative' model were correct, then vagotomy would be predicted to convert the response to loading of the same medullary neurone from a facilitation to an inhibition. Note also, however, that, due to lack of numbers, the effects of vagotomy upon the response to an elastic load (see Table D.1) are less clear than when a resistive load was used and further experiments with elastic loading are required.

As far as respiratory mechanics are concerned, there is a distinction between resistive and elastic loading; resistive loading is effective only during air movement but elastic loading is most effective at the end of an inspiration before expiration ensues. Therefore the time-course of facilitation or inhibition of medullary neurones should differ: resistive loading would be greatest at mid-inspiration and elastic loading at the end. This could be assessed by examining the time-course of neuronal discharge in more detail. It would, of course, require the time-base of the traces to be expanded.

To determine the pathways by which the afferent information reaches the medulla, experiments involving removal, recording and stimulation techniques could be devised as follows:

(1) Remove the neural inputs close to the proposed receptors either by sectioning or reversibly by cooling or drug-induced blocking.
(2) Record from the afferent fibres to establish the pattern of discharge during inspiratory loading; this should differ from inspiration in the absence of loading.
(3) Section the afferent fibres and stimulating them in a way that mimics the patterns in (2) above. The reflex responses should be duplicated in the absence of loading if the pattern of stimulation mimics that found during loading; if the pattern of stimulation mimics an unloaded breath, medullary neurones should fire normally.

The sites for sectioning, recording, etc. would be afferent vagal fibres close to the bronchiolar tree for experiments involving resistive loading and group Ib afferents in the case of the non-vagal input when there is elastic loading.

Two further comments are appropriate. Firstly, the experiments described above are technically extremely difficult to perform. This demonstrates that limitations in technique often restrict the experimental testing of hypotheses. Alternative, less difficult methods might be found, for example, dealing with the whole vagus nerve or spinocerebellar tract, but this will introduce interpretive problems as fibres with other effects might have been affected also. Secondly, the whole approach — both in the abstract and in the experiments suggested above — has been 'isolationist', considering the role of individual inputs or pathways. If medullary neurones receive converging inputs from many different sources, or if the inputs to spinal motor units come from more than one medullary cell responding in more than one way, then experimental results will be very complex to interpret.

Abstract E

The effect of occlusion of the common carotid arteries (carotid occlusion) on the firing pattern of hypothalamic supraoptic neurones

Introduction

Studies in female rats have shown that cells in the paraventricular nucleus of the hypothalamus can be divided into two types depending upon their patterns of discharge. The first group of cells, termed 'phasic neurones', shows a discharge pattern consisting of intermittent bursts of activity whereas the second group of cells, the 'random neurones', discharges in a more-or-less continuous fashion. In subsequent studies on neurones in both the paraventricular and supraoptic nuclei in the hypothalamus it has been shown that the reflex liberation of oxytocin is preceded by a brief acceleration in the firing rate of the random but not the phasic neurones. Therefore it has been suggested that the random neurones may be responsible for the release of oxytocin.

As yet, it has not been possible to show any correlation between the firing of the phasic neurones and the release of vasopressin. This is because many of the stimuli used to release vasopressin — such as an increase in the osmolality of the blood — will also release oxytocin. The present study was carried out in an attempt to correlate the firing of the phasic neurones in the supraoptic nucleus with the release of vasopressin. Occlusion of the common carotid arteries was used as the stimulus as this is known to release vasopressin preferentially.

Methods

Experiments were performed on female rats anaesthetized with urethane (25 per cent w/v; $0.6\,\mu l \cdot 100g$ rat^{-1} intraperitoneally). The discharge from neurones in the supraoptic nucleus was recorded extracellularly with glass micropipettes. Neurones in the supreoptic nucleus were identified by recording the electrical activity in the nucleus due to antidromic invasion of the cell bodies following electrical stimulation of the pituitary stalk.

Occlusion of the common carotid arteries was produced by placing small clamps on the arteries which closed the arteries without exerting any tension on the carotid sinus (which would have stimulated the baroreceptors).

In some lactating rats, intramammary pressure was recorded. This measurement can be used as a biological assay of oxytocin and vasopressin since both hormones are able to produce ejection of milk. However, at equal concentrations the activity of vasopressin is only 20–25 per cent of that shown by oxytocin.

Results

Of the neurones investigated, 33 were phasic, 60 were random and eight were 'silent', that is, they did not fire spontaneously. Phasic neurones exhibited intermittent bursts of activity alternating every 30–60 s with periods of silence. The periods between bursts of activity were either completely silent or contained a few isolated action potentials (Figure E.1). Random neurones showed more-or-less continuous discharge without any obvious pattern (see Figure E.2). Once a neurone had been allowed to stabilize for 5–10 min, the firing pattern did not change from phasic to random or in the opposite direction during the period of experimental occlusions.

There was a clear difference between the responses of phasic and random neurones to carotid occlusion. Figure E.1 illustrates three responses to bilateral carotid occlusion of a phasic neurone; Figure E.2 shows representative responses of three different random neurones to bilateral carotid occlusion and of the associated changes in intra-mammary pressure. When the phasic and random neurones were recorded from the same animal, the difference in their response to carotid occlusion was still observed.

Discussion

There is a number of problems associated with the interpretation of data obtained in these types of experiments. Firstly, when evaluating the responses of a phasic neurone one has to decide whether any burst after the stimulus resulted from it or whether the burst would have occurred anyway.

Secondly, when the extracellular activity of a neurone is recorded during a stimulus such as bilateral carotid occlusion, there is always

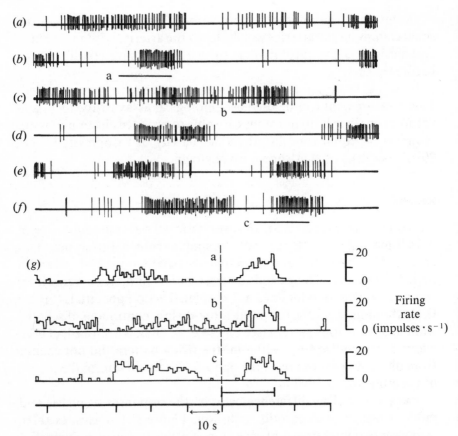

Fig. E.1 (*a*)–(*f*) Continuous record of spontaneous activity and bursts evoked by carotid occlusion in a phasic supraoptic neurone. Bars a, b and c mark three periods of occlusion of both common carotid arteries, each of 15 s duration: (*g*) the rates of firing of the neurone just before and after the three occlusions.

the possibility that any effect results from changes in blood pressure and local blood flow or is due to hypoxaemia, etc.

Fig. E.2 (opposite) Effects of bilateral carotid occlusion on random neurones. The rates of firing (records (*a*)–(*c*)) are from three different neurones. Time of occlusion shown by horizontal black bar. A milk-ejection response is shown in the upper part of each panel. These records of intra-mammary pressure were recorded at variable gains and are therefore not calibrated.

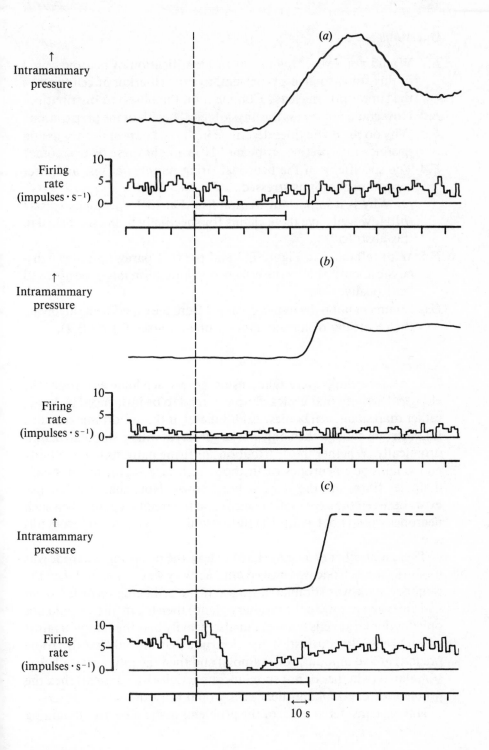

(a)

↑
Intramammary
pressure

Firing
rate
(impulses · s⁻¹)

(b)

↑
Intramammary
pressure

Firing
rate
(impulses · s⁻¹)

(c)

↑
Intramammary
pressure

Firing
rate
(impulses · s⁻¹)

10 s

Questions

E.1 Would you expect the means of identification of neurones used in this study to give a representative distribution of cell types in the supraoptic nucleus? Comment on the observed distribution.

E.2 How could you assess the physiological health of the preparation?

E.3 Why do the difficulties described in the last part of the discussion produce interpretive problems? How might these be overcome?

E.4 The specificity of the neuronal firing patterns at rest and after carotid occlusion is stressed.

(a) Why is it important to establish that such specificity exists?

(b) How well have these claims for specificity been established in this abstract?

E.5 With reference to Figure E.1 and the first paragraph of the discussion, outline how the effects of stimulation might be assessed statistically.

E.6 Comment upon the results obtained from the experiments involving the recording of intramammary pressure (see Figure E.2).

Suggested answers

E.1. As recordings were made using an extracellular electrode, the electrical activity that is picked up will tend to be influenced by large, rather than small, cell bodies. Additionally, if the antidromic stimulus is not strong enough then only the larger fibres will be stimulated antidromically. Whether the distributions of firing patterns in large-body and small-body neurones or in those with large-diameter or small-diameter fibres are the same is not known. Note that if a synapse exists between the supraoptic nucleus and the pituitary stalk, then such neurones would not be identifiable by antidromic stimulation from this site.

There is another problem related to how the recording electrode was used. If the electrode was moved until activity was found and then the pituitary stalk was stimulated (to see if the cell being recorded from had an axon passing to the pituitary gland) then the method would not only favour large cells (above) but also heavily bias the sample against 'silent' cells. By contrast, if the electrode was moved first (by some predetermined amount, for example) and then the pituitary stalk was stimulated (whether or not spontaneous activity was present) then the problem of overlooking silent cells would not exist.

Having considered some of the problems associated with obtaining a random sample of neurones from the supraoptic nucleus, we then should consider the distribution observed in the present experiments. The conventional view is that there are two populations of neurone in the supraoptic nucleus, which secrete vasopressin and oxytocin respectively (see introduction). The basis of the present experiments was an attempt to correlate one of these populations with a particular pattern of firing activity. Therefore, the observed distribution poses two problems. The first is the different numbers of random and phasic neurones. This distribution — 33 phasic and 60 random — is significantly different from an equal one ($P < 0.01$, binomial distribution). Assuming that the division between phasic and random neurones is not an artefact of identification (above) then a number of questions arises. These cannot readily be answered by the present protocol but they might give valuable information about the roles of these different neurones. Some of these questions are:

(1) Would the result be the same in males and females?
(2) Would the result be the same in virgin, pregnant, lactating or post-reproductive etc. females? (The present study on some lactating females did not give results relevant to this point.)
(3) Would the result differ in rats with hereditary diabetes insipidus (that is, animals which do not produce vasopressin)?

The second problem relates to the silent neurones, the third subgroup found in this study. Do they have a separate neurosecretory role and, if so, what is the substance that they secrete? Or are they some sort of 'inactive' group that can act if necessary as a reserve for phasic and random neurones? The present protocol does not address itself to these issues.

E.2. As well as controlling 'standard' variables (such as blood pressure, blood gases, body temperature) there is particular interest in establishing that blood volume and plasma osmolality are normal and the posterior pituitary is physiologically normal. Plasma osmolality and fluid volumes must be strictly controlled as these will affect spontaneous vasopressin release. This must be achieved in spite of the added problem of dehydration arising from surgery and lack of fluid intake. The difficulties are ameliorated somewhat by the experimental design since paired changes from prestimulus (control) values are being considered. However, individual animals can be compared more easily, both in the physiological and statistical sense, if the treatment of animals is standardized before and during the experiments. Alternatively, the effect of *systematically* changing the degree of hydration could be studied, but this would be a different investigation.

Other ways of testing the health of the preparation with particular reference to the present investigation would be to see if reflex responses to stimulation of the osmoreceptors or nipples produced 'normal' results. A complicating factor is that what is 'normal' under anaesthesia might not be known. Indeed, even the firing patterns of cells of the supraoptic nucleus and their responses to different stimuli might be abnormal under anaesthesia.

E.3. The problem is that it is the blood supplying the supraoptic nucleus that is important but the blood samples that are taken to help assess the physiological health of the preparation (see question E.2) are unlikely to have been taken from blood vessels supplying the supraoptic nucleus. To the extent that the hypothalamic blood supply comes from the internal carotid arteries, there is a real possibility that occlusion of the common carotid vessels will cause the type of problem described in the discussion. Moreover, there is very little reflex control of the blood supply to the brain under normal circumstances. The only real test of the physiological health of the area under study — and this is technically not possible at the present time — is to measure the gas composition of the brain extracellular fluid in the region of the supraoptic nucleus. However, what is technically possible is to measure regional blood flow by considering the rate of loss of heat or radioactive materials (for example, ^{86}Rb), from the blood before and during occlusion of both carotid arteries. Such experiments might be very difficult to perform upon the animals from which electrical activity is being recorded and this would argue for preliminary or 'pilot' experiments to be performed to clarify this problem. However, inter-individual variability in the anatomy of the circle of Willis might produce considerable variation in the effects of bilateral carotid occlusion; this would argue for experimental determination of blood flow *in each experiment* in preference to separate pilot studies.

If a variable such as the concentration of oxygen in the region of the supraoptic nuclei were to change, it is easy to speculate why the firing pattern should also change. Thus, one could argue in the early stages that the ischaemic cells would lose potassium and so brain extracellular potassium would rise and lower cell membrane potential and firing threshold; later the cells might become less excitable as membrane potential continued to fall. The important point, of course, is that any result obtained in such circumstances would provide interpretive problems.

E.4. The division of neurones into two types on the basis of their firing patterns is sensible only if these patterns are found in all animals and if there is a consistent pattern in any single neurone. The proposal to investigate the correlation between firing pattern and the type of transmitter released would be nonsense if examples could be found where the firing pattern in any neurone changed with the passage of time. It would also have been unsatisfactory if there were cases where the firing pattern did not clearly fall into either of the two categories (this ignores the 'silent' neurones, of course).

When the response of the neurones to carotid occlusion is considered it is quite impossible to make a judgement based on the information in the abstract. Thus, Figure E.1 shows the increase in firing produced in only one neurone, and Figure E.2 shows changes produced in only three! To claim that there is a specific response to stimulation requires it to be shown that, for a whole group of cells — in other words, for a statistical sample — this response depends upon whether the resting pattern was phasic or random. (The assessment of the response will be discussed in question E.5). The distribution of responses (as 'increase' versus 'no increase') between 'phasic' and 'random' cells can then be tested by the χ^2 test, the null hypothesis being that there is no difference between the responses of the two cell types. Such an analysis also requires that the stimulus used in all cases was the same; this does not seem to be the case. Thus, whereas the single 'phasic' neurone was stimulated for 15 s (see Figure E.1), the 'random' neurones (see Figure E.2) were stimulated for different periods of time (35–60 s).

It would have been informative to have studied also the effect upon each cell type of a stimulus believed to stimulate random neurones only — the effects of nipple stimulation (see introduction). This would have enabled a comparison to be made for each neurone between the effects of two stimuli, both of which were believed to be specific in the type of cell that they influenced. The implication of the authors' hypothesis is that a cell would respond only to one of the two stimuli, and that its firing pattern at rest would determine which one.

Finally, no information is given with regard to the responses of the 'silent' neurones.

E.5. Observation of Figure E.1 certainly suggests that stimulation had an effect on the firing pattern. The difficulty is to assess any such effect quantitatively. This problem is exacerbated in the case of phasically firing neurones since excitation of them could:

(1) prolong a burst of activity and shorten the gaps between bursts;
(2) increase activity during a burst (but not prolong it);
(3) increase 'basal activity' between bursts; or
(4) cause a mixture of these changes.

Quite possibly any of these effects could depend upon the time in the firing/silent cycle when the stimulus was applied. An assessment of this kind of neurophysiolgical problem is a specialized topic and so outside the scope of the present book. However, it is relevant to the scientific method in general to consider in outline how these problems could be approached. One could investigate some of the possibilities by measuring:

(1) the length of a burst of activity (or number of impulses per burst) after stimulating the neurone during a burst;
(2) the length of a 'silent period' (or number of impulses per 'silent period') after stimulating the neurone after a burst.

In both cases, the result would be compared with that obtained in the absence of stimulation (or during stimulation by an oxytocin-releasing stimulus).

With respect to the observation that each cell was tested a number of times, we reiterate that, statistically, if ten different cells are each stimulated three times, then the sample size is ten, not 30. The advantage of stimulating any cell more than once is that the experimenter can be more confident that the average result (increase or no increase) is not a freak one. In addition, it should be noted that if responses are recorded from a number of cells from each of a number of animals then analysis of variance techniques become appropriate. These could be used to compare the variances due to animals and cell type or, if the same cell was used with different stimuli, those due to animals and stimulus type.

E.6. A major problem that has already been considered (question E.4) is that the stimulus as used in Figure E.2 is not of a constant length and never equalled the (constant) value shown in Figure E.1 (clearly there is no way of knowing if the same stimulus length was used for different phasic neurones). Further, the intramammary record of pressure is of no quantitative use because no calibration is given. Even so, the record in trial C is obviously at too large a gain (as it goes off the scale) and that in case A has oscillations superimposed (1 per 6 s), the origin in which is obscure.

However, the *time-course* of the pressure record might be of use. Thus, the general premise is that rises in intramammary pressure can act as a bioassay method for vasopressin release because this hormone shows a weak milk-ejection activity. Therefore, it could be argued that, after a delay following carotid occlusion and an increase in firing of phasic neurones, there will be an increase in vasopressin circulating in the blood and, hence, an increase in intramammary pressure. Moreover, since there would be no increase in firing of random neurones there would not be an increase in the release of oxytocin. Presumably, a stimulus such as nipple stimulation would increase the firing of random (but not phasic) neurones and also raise intramammary pressure faster and/or by a greater amount.

It might seem that evidence in favour of this argument exists when the results shown in Figure E.2 are examined. Thus there is no general increase in firing of random neurones produced by the stimulus but there is an increase in intramammary pressure. However, this might be a false conclusion to draw. The experimenter must disprove alternative interpretations of these results. Two alternative explanations are:

(1) The stimulus is not specific to phasic neurones. Thus the response to carotid occlusion of the third neurone (Figure E.2c) could be described as 'a brief acceleration in the firing rate', a description applied to the neural event preceding the release of oxytocin (see introduction).

(2) Phasic neurones (or at least some of them) release oxytocin. The rise in intramammary pressure could be due to circulating oxytocin as well as (or instead of) vasopressin (see methods section).

These alternatives need to be eliminated and indicate the problems of interpretation that arise when aspects of the protocol and assay are inadequate. In particular, it would seem that the development of a specific assay system that distinguishes between oxytocin and vasopressin (the bioassay system used here was unsatisfactory in this respect) is required.

It is worth concluding by stating that the different roles of phasic and random neurones when vasopressin is released under physiological circumstances remain to be determined.

Effects of drink temperature on the emptying of liquid from the human stomach

Introduction

The physiology of gastric emptying can be studied in man by non-invasive techniques. It has been found that the emptying of liquid from the stomach can be divided into two phases: an initial, rapid phase followed by a slower exponential decline. The initial rapid phase of emptying can be inferred from the '5 min volume', that is, the amount of fluid remaining 5 min after fluid ingestion. The subsequent slower phase can be described in terms of its half-life and is calculated from a regression analysis of the relationship between the time since the end of the rapid phase and the logarithm of the volume of fluid remaining in the stomach. However no details are known of the factors that might influence these processes.

The series of experiments described here was designed to investigate the effects of temperature of the ingested fluid on emptying in a group of normal volunteer subjects. The technique used to assess the volume of fluid in the stomach is based on ultrasound scanning.

Methods

Eight normal male volunteer subjects aged 21–29 years participated in the experiments. They attended the laboratory at 10 a.m. after an overnight fast and were seated in an upright chair for 5 min before each drink and throughout the experimental period. The drink given was 500 ml of a proprietary cordial and its strength was standardized at a value that was palatable to the volunteers. Drinks were given at temperatures of either 12 °C or 37 °C. The order in which the drinks were given was randomized, and they were consumed at a speed which was as rapid as the subjects could comfortably manage (mean of 38 s).

Gastric volume was calculated from real-time ultrasound measurements made before each drink and at 5, 10, 15, 20, 25, 30, 40, 50 and 60 min afterwards. The average of two measurements at each time point was used in the calculation of gastric volume. The method for estimating gastric volume is based on the fact that the stomach transmits sound when containing liquid and thus its walls can be clearly outlined by an ultrasound scanner. The subject held his breath and then a scan was made with a specially designed probe while traversing the width of the stomach at right angles to its long axis. A microcomputer was used to calculate the cross-sectional area of each 'slice' from the measurements of the distance between the walls of the stomach at different stages of the traverse. The total volume of the stomach was then calculated by summation of the measurements from a series of regularly spaced (1 cm apart) cross-sections of the stomach.

In vitro measurements have shown that this is an extremely accurate and reproducible method.

Results

The results are summarized in Table F.1. They indicate that:

(1) the volume of liquid in the stomach 5 min after ingestion was significantly less in the case of the colder drink ($P < 0.01$, Student's paired t test);

(2) the half-life of emptying of the stomach during the slow phase did not depend upon the temperature of the drink ($P = 0.87$, Student's paired t test); and that

(3) there was a direct linear relationship between the '5 min volume' of the stomach (x) and the subsequent half-life of emptying during the slow phase (y). For the drink at 12 °C, the correlation coefficient, r, was 0.87 and the equation of the regression line was:

$$- y = 0.08 \cdot x - 5.1$$

For the drink at 37 °C, the correlation coefficient was 0.79 and the equation of the regression line was:

$$y = 0.09 \cdot x - 11.1$$

Table F.1 '5 min volume' (ml) and half-life of emptying during the slow
phase (min) for drink at 12 °C and 37 °C. Results from eight subjects (A–H)

	12 °C		37 °C	
Subject	5 Min volume	Half-life	5 Min volume	Half-life
A	376	28.0	428	24.8
B	320	26.9	320	16.8
C	367	17.7	394	22.1
D	166	11.0	289	11.0
E	133	3.8	228	11.8
F	242	10.1	278	6.7
G	184	11.7	220	4.7
H	209	9.8	300	24.9
Mean	250	14.8	307	15.3
± SE	33.1	3.0	25.8	2.8

Questions

F.1 What do you understand by the concept of a 'physiologically
healthy preparation' in experiments such as these?

F.2 Comment upon the protocol.

F.3 How adequate do you consider the stimulus to have been?

F.4 Outline how the calculation of stomach volume is performed and
comment upon the assumptions that are made.

F.5 The results are presented as three findings.
Comment upon these and include statistical comment where
appropriate.

F.6 Outline some of the difficulties that might be associated with
further experimental work.

Suggested Answers
Note
The use of human volunteers will require previous acceptance of the
protocol by an Ethics Committee and a complete explanation of it to
the volunteers. One advantage of using human volunteers rather than
animals is that the subjects can be asked to ingest the drink as quickly
as possible and this is an important part of the present protocol. In
addition, stress (which might alter gastric function) can be minimized

in human volunteers if they are reassured by the experimenters. With proper training, animals such as dogs would also be unstressed by the procedures used here, but it might take longer to make them thoroughly familiar with the protocol.

F.1. The concept of assessing a 'physiologically healthy preparation' can be rather different with human experiments. The continuous monitoring of many 'normal' variables (PCO_2, blood osmotic pressure, etc.) is not generally possible as this requires invasive techniques, but non-invasive means could be used for measurements of blood pressure, pulse and respiratory frequency. More elaborate techniques (giving urine samples, measuring reflexes and recording the e.e.g., for example) are all possible, with subject consent of course. However, what will rapidly become apparent to the experimenters when conscious human volunteers are used is the lability of measurements within an individual and the extent of interindividual variation. This emphasizes the importances of eliminating stress, keeping subjects rested, pre-experimental controls, etc. and the advantages of a paired experimental design (as in this study).

In most experiments involving humans, the volunteers are required to have passed a medical examination before selection for the experiments. This gives the experimenters the opportunity to check for clinical abnormalities and 'unusual' medical histories. In these particular experiments, special attention should be given to asking volunteers about their dietary habits and screening for evidence of gastrointestinal disorders. Subjects with unusual diets or gastrointestinal disorders should be excluded if 'healthy' volunteers are required but, in other studies, such volunteers — when compared with healthy controls — might be of *most* interest.

F.2. Ingestion of fluid took place at the same time each day, a convenient arrangement as well as one which takes into account (but does not investigate) circadian changes. The control of posture is commendable, not only because changes in posture might alter emptying of the stomach (possibly via effects on abdominal pressure) but also because they might result in changes in the shape of the stomach. This is an important consideration when the means for calculating the volume of the stomach by an algorithm is considered. Thus the mathematical model might estimate the volume of some gastric shapes more accurately than others (see F.4, below).

The previous 'overnight fast' is designed to ensure that the stomach is initially empty. No evidence is presented to indicate that this objective was achieved; even if it were not, to describe the volume of fluid in the stomach in terms of *increments* over this resting value would have been a possible solution. This issue highlights a difference between some animal and human experiments. A longer fasting period — together with a controlled diet for some days before this — would be inconvenient to human volunteers (unless they were in a hospital ward) but it is easier to administer as part of the daily attention that an animal receives.

The randomized sequence of warm and cold drinks is desirable as a principle, of course. Successive experiments can be performed as long as there is no 'carry-over' from the previous experiments. Physiologically, the next day will be acceptable in this regard; however, the timing of experiments with humans can be determined very much by the wishes of the volunteers! It would have been desirable to have repeated in-gestion of a drink at the first temperature at the end of the experiment. This would require the experiment to last at least one more day, of course, but this last stage is a kind of 'post-experimental control'; by comparing the results from it with those from the first stage (when the temperature of the drink is the same), we would have the opportunity to examine the consistency between results. If results from stages 1 and 3 do differ, we might be able to use them to better estimate the 'baseline' for phase 2 or even to develop a hypothesis to account for the differences and so to understand better the process of stomach emptying.

F.3. The use of a palatable drink of a proprietary brand of orange cordial would certainly seem to be a 'physiological' stimulus in some respects. The composition of the drink is not given in any detail even though we are told that it is constant throughout the experimental series. If a comparison with the results of others is required, then the same drink must be used, its composition being obtained from the scientific literature. (Details of the composition of the drink would be required in a published paper so that others could repeat the experiments). By contrast, the use of a drink with a composition *different from* that used by others might be valuable in establishing the effect of *changes* in composition, etc. on the '5 min volume' or slow phase of emptying. Of particular interest would be the effects of pH, glucose concentration and total osmolality as well as the effects due to the simultaneous presence of alcohol, caffeine, lipid and non-digestible material such as cellulose. These latter substances seem particularly important since the stomach is generally regarded as a reservoir of *food* rather than *drink*. Also, most meals are a mixture of solid food as well as drink. In this respect, the stimulus was slightly abnormal. It is a reflection of the general principle that the stimulus in an experiment is usually 'simplified' when compared with the natural one. As has been discussed in Part I of the book, this solves one set of interpretive problems and raises another!

There are two other problems with the experimental design. First, the ingestion of the drink 'as rapidly as the subjects could comfortably manage' has the advantage that it enables the starting-time to be defined more accurately; however, it is almost certainly neither a socially acceptable stimulus nor the usual physiological one! Normal stimuli to the stomach are likely to be far less brutal in this respect than those in the present study. Second, and potentially more serious, the difference in temperature between the drinks is unlikely to be maintained throughout the whole of the 60 min period of experimental measurement. The possible loss of the difference in temperature between the stimuli poses interpretative difficulties. Consider the possibility that no difference between the effect of the drinks ingested at the two temperatures is found. This might indicate a lack of effect of temperature but might also indicate the loss of a difference between the temperature of the two drinks. Conversely, when a difference between the two drinks is found, it might underestimate the true effect of temperature as the stimulus was so transient. Recordings of the temperature of the contents of the stomach are required; in the absence of these, there must be scepticism that the intended experimental stimulus was, in fact, given and doubts as to what exactly was responsible for any differences that were observed.

F.4. The stages of calculation are as follows:

(i) *Calculation of the area of each 'slice' through the stomach.* This requires a knowledge of the distance between the walls of the stomach (d_x) at a series of points during a traverse of the stomach by the scanner. The distance can be assessed from the delay recorded by the scanner between the echoes from the front and back walls of the stomach. The area of the 'slice' can then be computed by means of an algorithm which estimates area from a series of values of d_x taken at known points during the traverse. Thus:

$$\text{Area} = \sum_{\substack{\text{end of traverse} \\ \text{start of traverse}}} (d_x \cdot S_x) \quad \begin{array}{l} \text{where } S_x \text{ is the distance} \\ \text{moved by the scanner} \\ \text{between successive} \\ \text{measurements of } d_x \end{array}$$

(ii) *Calculation of the volume of each 'slice'*. The easiest possibility is to assume that it is 1 cm thick (the distance between successive traverses by the scanner) and to assume that the scan was made through the centre of this 'slice'. Another possibility is to calculate the volume between adjacent cross-sections, that is to assume that the scanner passes along the boundary of each 'slice'.

(iii) *Calculate the volume of all 'slices'*. This is a simple addition of the results from (ii) above for all traverses of the stomach. Assumptions are obviously made in stages (i) and (ii) above. Technically, the accuracy of the method depends critically upon being able to locate exactly the walls of the stomach. When the walls are not smooth or cannot easily be scanned then some doubt as to the accuracy must arise. (This has been the case with similar investigations of heart volume or fetal bladder volume, for example.)

When the calculation is considered, then some assumption as to the shape of the stomach has to be made. In the first possibility described in (ii) above, it was that the stomach was equivalent to a series of rectangular blocks. In other cases it would be normal to assume that, between two adjacent values (whether these are readings of the distances between the front and back walls of the stomach or the calculated areas of cross-sections as in the second possibility in (ii) above), there is a regular change so that intermediate values can be calculated by inter-polation. In these cases the assumption would then have to be made that one were dealing with a truncated cone or rectangular pyramid or some other solid so that the volume could be calculated from the measurements available. This need not be so. Moreover, the errors so introduced might depend upon the overall shape of the stomach which might alter with body posture or total gastric volume. (However, the protocol in these experiments does control these two variables as much as possible and the effects of interindividual variation are reduced by use of a paired experimental design.)

An independent test of the reliability of measuring stomach volume by this method is required. The text indicates that, *in vitro*, the technique gives results that 'accurately and reproducibly' agree with results obtained by direct measurement. However, the accuracy of the method *in vivo* remains unknown. One check would be to compare results from the present method with those obtained by aspiration of the contents of the stomach. Again, such a comparison requires acceptance by volunteers; it is also an example of the fairly common procedure of having to 'calibrate' a newer, more convenient technique against an older, less convenient, 'standard' or direct technique.

F.5. The duplicate measurement of all results does not enable the sample size to be doubled; instead it means that the experimenters can place greater reliance on an average value since it will be less affected by a 'freak' result than will a single measurement. In practice, if the two results differed markedly, a third value should be obtained *during the experiment* to enable one of the first two to be discarded. Clearly, for this reason, there is an advantage in being able to calculate results during, rather than after, the experiment.

Result 1. When the data (Table F.1) are analysed by Student's paired *t* test, the result is:

$$t = 3.93; \text{d.f.} = 7, P < 0.01$$

This indicates a significant difference between the '5 min volume' at the two temperatures.

Incidentally, had the two sets of data been compared by Student's unpaired *t* test (that is, treating them as coming from two independent groups) then a non-significant difference between them would have been found. This is a type II error and illustrates the strength of the paired design of the experiment.

Result 2. The half-lives of the slow phases of emptying are not significantly different when analysed by Student's paired *t* tests. If we can assume that a type II error has not been made — that is, that a small difference in half-lives has not been 'missed' — then the result implies that the two drinks are emptied at *different* rates when these are expressed not as a fraction of the '5 min volume' of the stomach per unit time but rather as an absolute volume (measured in ml) per unit time. Table F.1 shows that, at 5 min, approximately 250 ml of fluid remain in the stomach after ingestion of a drink at 12 °C and 300 ml when the fluid was at 37 °C. If there is equality of half-lives during the slow phase of emptying (say 15 min) then the volumes of fluid expelled in the first four 15 min intervals of the slow phase of emptying (that is 5−20, 20−35, 35−50 and 50−65 min after ingestion of the drink) would be:

at 12 °C, 125 ml, 62 ml, 31 ml, 16 ml, and
at 37 °C, 150 ml, 75 ml, 38 ml, 19 ml.

This analysis is slightly unusual in that it describes the process of gastric emptying in terms of two separated phases. The initial phase is finished at 5 min and is then succeeded by the slow phase. Normally the two phases would overlap and the analysis would take this into account. Thus, to calculate the half-life of the slow phase of emptying, zero time would be taken as the time of ingestion (not 5 min as here) and the starting volume would be estimated by extrapolation to zero time of the curve relating gastric volume to time (not the '5 min volume' as here).

Result 3. The third finding was that the half-lives during the slow phase of emptying were directly proportioned to the '5 min volumes' and that this result was obtained when ingestion of fluid at both 12 °C and 37 °C was considered. This suggests that the rates of emptying of fluid from the stomach (volume per unit time) at the two temperatures are not as different as indicated by the figures above. At first it might appear that there is an internal inconsistency between the three results. Thus it might be argued that:

Since a higher temperature of ingested fluid leads to a higher '5 min volume' (result 1);
and since the half-life of the slow phase of emptying is directly pro-portional to the '5 min volume' (result 3);
then the half-life of the slow phase of emptying would be greater with a higher temperature of ingested fluid.

This deduction is not the same as result 2 where the half-life was found to be independent of the temperature of the drink. As has already been described, there is a difficulty of interpretation in this study since we do not know for how long the different temperatures of the stimuli were maintained; therefore, the results relating to the effect of temperature on the slow phase of emptying (result 2) might be spurious anyway. Even so, and even if the temperatures were the same during the slow phases of emptying, the apparent inconsistency between the results requires an explanation.

One possible explanation (already mentioned) is that result 2 is an example of a type II error, that is, of having 'missed' a real difference. However there is another explanation. It is based on the fact that results 1 and 2 were obtained by analysis of paired data, comparing

the results of the two ingestion temperatures *for each individual.*
By contrast, the two analyses giving result 3 were both performed by
considering the individuals *as a group*, once with a drink at 12 °C and
then at 37 °C.

Figure F.1 explains the paradox in simplified form. The results are
to be considered as coming from four individuals, each performing the
experiment at the higher (○) and lower (●) temperatures. Lines join
the pairs of results for clarity. These results show:

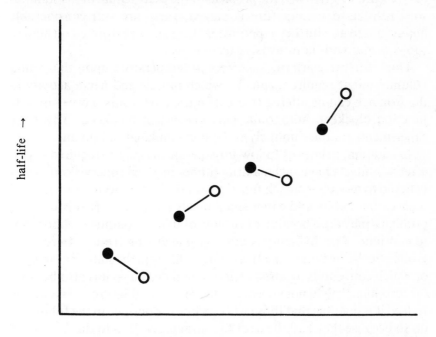

5 min volume →

Fig. F.1   Illustrates, in simplified form, the results obtained in the abstract;
●, drink at 12 °C; ○, drink at 37 °C. Results from same subjects are joined
by a line.

(1) For paired, but not unpaired data, the '5 min volumes' at 37 °C
    are greater than those at 12 °C (see result 1);
(2) there is no consistent difference between the half-lives at the two
    temperatures (see result 2); two are higher at 37 °C than 12 °C
    and two are lower; and
(3) for data obtained both with the drink at 37 °C and at 12 °C, the
    groups show a positive correlation between '5 min volume' and

half-life and the equations of the two regression lines would be similar (see result 3).

In fact, the data of Table F.1, when plotted in this way, show a very similar result to those in Figure F.1. This kind of problem has been discussed more fully in Chapter 6.

F.6. Even if the method used in the present experiments can be shown to be accurate *in vivo* and the problem of the exact nature of the stimulus with respect to temperature is solved, there are still considerable limitations to any further experiments that can be performed on human subjects and with non-invasive techniques.

Thus, further work on the effect of temperature upon the '5 min volume' might require means by which muscle and nerve activity in the stomach can be altered (? use of muscle relaxants, nerve–muscle junction blockers, autonomic nervous system blockers). These are experiments that are unlikely to be contemplated in humans.

In addition, a series of follow-up experiments may be required — say, after volunteers have taken a drug (under medical supervision) that is believed to modify stomach function or after the subjects have changed their eating habits and eaten small meals frequently. If this is done, problems may arise because of the loss of data if volunteers choose not to continue. The difficulty is not just one arising from a decrease in group size that makes a type II error more likely in the statistical analysis or which confounds a paired experimental design — it is also because the 'dropouts' might not be representative of the initial group as a whole. Indeed it is possible that those who decline to continue in the follow-up do so because they had the least favourable reaction to the first set of experiments and it is *for this very reason that they are the subjects in whom the experimenters are most interested!*

Index

(Pages in brackets refer to suggested answers to abstracts in Part II)

ablation 44, 45
acid-base balance 33−5, (184, 185)
alternative hypotheses 94, 174
anaesthetic
  effect of *see* anaesthetised *v.*
    conscious preparation
  level 14, 22, 32, 33
  type 14−16
anaesthetized *v.* conscious
    preparations 12, 14, 15, 18,
    19, 31, (184)
animal care (*see also* ethics;
    species) 29−31
animal models (*see also* species) 9, 10
anomalous results *see* hypotheses
ANOVA 23, 27, 83, 91, 93, (175,
    186)
apparatus *see* instrumentation
artefacts
  of correlation *see* causality;
    correlation; linear regression
  of recording *see* recording
  of stimulation *see* blockade;
    stimulation, adequacy of,
    specificity of
assays *see* biological assay;
    chemical assay

bias *see* samples
binomial test 83, 97
biological assay 53, 43, (157, 186,
  187)
biological models *see* modelling
blockade 43, 44
blood pressure 32−5, (159, 160)

calculations 76−81
  control values 76−9, (156, 157)
  experimental changes 24, 25,
    76−81, (132, 133, 143, 144)
  hypotheses and 79, 80
calibration (*see also*
    standardisation)
  of assays 53
  of instruments 48−53, (194, 195)
carbon dioxide *see* gas tensions
care of animals *see* animal care
categorisation of results 86, 89, 90,
    95, 97−100, (157−9, 171,
    172, 182, 183, 185)
causality (*see also* correlation)
    105−8, (135, 171−4, 185, 186)
cell studies *see in vitro, v. in vivo*
    preparations
chemical assay 53, 54, (131, 132)
  storage 53
chi-squared test 83, 97, (158, 159,
    175, 185)
Cochran Q test 83
coefficient of variation 47
compensatory mechanisms *see*
    homeostatic compensation

conscious v. anaesthetized
  preparations 12, 14, 15, 18,
  29, 31, (184)
contingency coefficient 83
control experiments (*see also* pilot
  experiments) 23–5, 27, 32,
  35, 36, 45, 77–9, 86, 101–3,
  (133, 143, 156, 157)
control groups 19–21, 24, 25, 86,
  101, 102, (143, 156, 157)
control phase (*see also* calculations)
  19, 21–4, 29–38, 45, 76, 77,
  86, 101, 102, (130, 133, 142,
  143, 156, 157, 171–3, 191, 192)
correlation (*see also* causality)
  63–6, 74, 83, 85, 86, 96, 97,
  105–8, 116, (134, 135)
cumulation frequency histogram 69

data collection 3, 8, 27
decerebrate preparation 14
decorticate preparation 14
documentation of results 46
double-blind study 25, 26
downgrading results 89, 90, 98,
  (158, 159, 174, 175)
drug studies 8, 13, 22, 24, 25, 36,
  40, 77, 79, 80, 91, (157)

epidemiology 4–7
ethics 9, 14, 26, (190–2, 198)
experimental design (*see also*
  control experiments; control
  groups; control phase;
  experimental group;
  experimental phase; pilot
  experiments; statistics) 19,
  21, 27, 28, 41–3, 71, 82–92,
  101–3, 112, 119, (133, 134,
  146, 147, 156, 157, 172, 173,
  185–7, 191, 192, 194)
experimental group 20, 21, 24, 25,
  39–57, 86, 101, 102, (156,
  157)
experimental phase (*see also*
  calculations; multiple

experimental phases) 21–4,
  32, 39–57, 76, 86, 101, 102,
  (156, 157, 172, 173)
experimental preparation (*see also*
  anaesthetised *v.* conscious
  preparations; health of
  preparation; *in vitro v. in vivo*
  preparations; levels of
  experimental preparation;
  preparation) 12, 14, 17, 18

Fisher exact test 83, 99, (175)
formulating hypotheses *see*
  hypotheses
frequency distribution 68–71
frequency response of apparatus
  49–53
Friedman analysis of variance 83

gas tensions 35, 36
graphs *see* presentation of results
grouped results, presentation of
  64–6, 68–72

health of preparation (*see also*
  blood pressure; control phase,
  heart rate; special health
  factors, etc.) 22, 29–38, 84,
  101–3, 110, 111, (130, 131,
  133, 156, 171, 183, 184, 190,
  191)
  in animals 30–8, (156, 171)
  in conscious humans 29, 30,
    (191)
  in *in vitro* experiments 17, 32,
    36, 37, (130, 131)
heart rate 32–5, (159, 160)
homeostatic compensation 12–14,
  17, 18, 36, 108–12, (158, 177)
hormones 21, 22, 24, 31, 32, 35,
  36, 102, 103
hypotheses
  anomalous results and 4, 16, 40,
    46, 72, 73, 111, (192, 196–8)
  calculations and 79, 80
  formulating 3–5, 18, 115, 119
  indecisive results and 112, (193,
    196–8)

presenting results and 58, 62, 63, 72–81, (173, 174, 186, 196–8)
refining, modifying 7, 14, 103–5, 110, 112–19, (136, 148, 160–2, 176, 177, 182, 183, 187, 198)
refuting 6, 7, 72, 108–11
statistics and 19, 82–4, 86, (144–7, 186)
supporting 103–5, (187)
testing predictions and 3, 5–8, 72, 100, 103, 117, (136, 146–8, 162, 167–70, 176, 177, 187, 198)

individual results, presentation of 63, 64, 66–8, 71, 72, (173, 174, 186)
instrumentation 46–53, 122
accuracy of 46–9, 82
calibration of 48–53, (194, 195)
frequency response of 49–53
precision of 46, 47
stability of 49–52
technical limitations and 57, (173, 177)
integrated responses *see* homeostatic compensation
interval data 87, 89, 90, 96, 97, 99, 100
in vitro v. in vivo preparations (*see also* levels of experimental preparation) 12–14, 16–18, 30–7, 40, 41, 109, 110, (129, 130, 145, 146)

Kendall rank correlation coefficient 83
Kolmogorov-Smirnov test 69, 83, (149)
Kruskal-Wallis test 83

levels of experimental preparation (*see also in vitro v. in vivo* preparations) 12, 14, 17, 18

levels of statistical measurement (*see also* interval data; nominal data; ordinal data) 19, 86–92, (149)
linear regression 72, 73, 80, 81, 116
logarithms *see* mathematical transformation of data
longitudinal studies 26, 27, 86, (198)

McNemar test 83, 97
man *see* species
Mann-Whitney test 83, 99, (158, 159)
matching samples *see* samples
mathematical models *see* modelling
mathematical transformation of data 58–61, 64, 68, 88, 116, 117, (174)
mean 87–9
median 87–90
mode 87, 89, 90
modelling
biological models (*see also* species) 9, (160, 161, 167–71)
mathematical models, physical models 58–61, 116–19, (148, 149, 191, 193–5)
modifying hypotheses *see* hypotheses
movement artefacts *see* recording
multiple experimental phases 21–3, 66, 86, (133)

nominal data (*see also* categorisation of results) 86, 89, 95, 97–100
normal distribution 83, 87, 88
normalising data *see* mathematical transformation of data
null hypothesis 93, 94, (146, 147)
number of experiments 19, 46, 85, 89, 95–100, (173, 185)

ogive 69
one-tailed statistical test 94, 98, (132)

ordering of experimental phases *see* multiple experimental phases
ordinal data 87, 89, 96, 97, 99, 100
organ studies *see in vitro v. in vivo* preparations
osmotic pressure 35, 36, (183, 184)
oxygen *see* gas tensions

paired data 91–2, (136, 174–6, 195)
paired experimental design *see* presentation of results; statistics
pairing samples *see* samples
Pearson product-moment correlation coefficient (*see also* correlation) 83
percentiles 87–9
pH *see* acid-base balance
physical models *see* modelling
pilot experiments 35, 36, 53, 54, (156, 171, 184)
placebo 25, 26
Poisson distribution (148)
precision of apparatus *see* instrumentation
predictions of hypotheses *see* hypotheses
preparation (*see also* anaesthetized *v.* conscious preparations; health of preparation; *in vitro v. in vivo* preparations; levels of experimental preparation)
  decerebrate 14
  decorticate 14
  spinal 14
presentation of results 55, 58–81
  correlation studies 63–6
  frequency distribution 68–71
  grouped results 64–6, 68–72
  individual results 63, 64, 66–8, 71, 72, (173, 174, 186)
  mathematical transformation of results 58–61, 64, 68, 116, 117
  paired results 66–8, (196–8)
  scales of graphs 63, 64, 74–6, (144, 176, 177, 186, 187)

time-course of results 71, 72, 74, 107, 108, (144, 145, 176, 177, 187)
type of hypothesis 58, 62, 63, 72–81, (173, 174, 186, 196–8)
units 58, 61–3, (132, 133, 143, 144)
unpaired results 68, 69, (196–8)
P values 93, 94

quartiles 87–9

random samples *see* samples
range 87
recording (*see also* instrumentation) 46–57, 82, (171, 172, 186, 187)
  artefacts in general 55–7, (172, 186, 187)
  artefacts due to movement 55, 56, (172)
  artefacts of selectivity 56, 57, (182, 183)
  documentation 46
  standardisation 54, 55, (148, 157, 158, 171, 172, 182, 185, 187, 195)
redundancy *see* homeostatic compensation
refining hypotheses *see* hypotheses
refuting hypotheses *see* hypotheses
regression *see* linear regression
replacement *see* ablation
response specificity *see* specificity
resting values *see* control phase
results *see* calculations; categorisation of results; homeostatic compensation; hypotheses; presentation of results; recording; statistics)

samples
  biased *v.* random 19, 20, 26, 27, 56, 57, (132, 133, 149, 157, 158, 182, 183, 191, 198)
  criteria for categorisation into nominal format *see*

categorisation of results
criteria for matching (pairing) of
7, 20, 21, 86, 91, 104, 105,
107
paired (matched) *v.* unpaired
(unmatched) 20, 21, 26, 27,
66–9, 78, 91, 92, (136, 172,
183, 185, 191, 192)
scales 63, 64, 74–6, (144, 176, 177,
186, 187)
signal/noise ratio 76
single-blind study 25, 26
Spearman rank correlation
coefficient 83, 96, 97
special health factors (*see also*
health of preparation) 35, 36,
(130, 131, 156, 171, 183, 184,
191)
species 8, 12, (142, 143, 183)
animals as models for man 9, 10
extrapolation between species
8–11, 18, (134, 136)
factors influencing choice of
9–12
man (*see also* subjective
measurements) 8, 9, (190–2,
198)
specificity
of assay 54, (131, 132, 157, 187)
of blockade 43, 44
of response 14, 16, 37, 38,
101–3, (129, 130)
of stimulus 14, 16, 37, 39–42,
101–3, 109, 110, (129, 130,
134, 148, 156–8, 161, 162,
185, 187, 193, 196)
spinal preparation 14
stability
of instruments 49–52
of samples 53
standard deviation 87–9, 99, (132,
159, 160)
standard error 87–9, 99, (132,
159, 160)
standardisation
of assays 53
of blockade 44

of categories *see* categorisation
of results
of experimental conditions *see*
control phase
of instruments 48–53, (194,
195
of recording *see* recording
of stimulus *see* stimulation
statistics
(*for terms and tests, see
individual entries*)
choice of statistics tests 83–92
experimental design and 19,
82–4, 86, (144–7, 186)
level of measurement in 19,
86–92, (149)
number of experiments and 19,
46, 85, 89, 95–100, (173,
185)
paired *v.* unpaired statistics test
91, 92, (136, 174–6, 195)
testing hypotheses and 19,
82–4, 86, (144–7, 186)
type I and type II errors in 21,
84, 85, 93–100, (135, 195,
198)
stimulation, stimulus (*see also*
ablation) 39, 40, 109, 110
adequacy of 40, 41, (134, 161,
162, 185, 192, 193)
relation of, to normal
physiology 40–3, 108, 110,
111, (130, 136, 157, 158, 192,
193)
specificity of *see* specificity
standardisation of 39, 40, 109,
110, (157, 158, 183–6,
191–3)
storage of samples 53
stress *see* health of preparation
Student's paired t test 83, 91, 92,
(195)
Student's unpaired t test 83, 90, 99,
(132, 135, 136, 158, 159, 195)
subjective measurements 25, 26
supporting hypotheses *see*
hypotheses

technical limitations *see*
    instrumentation; hypotheses,
    refining
temperature 33, 36
testing hypotheses *see* hypotheses
time-course *see* presentation of
    results
tissue studies *see in vitro v. in vivo*
    preparations
transformation of data *see*
    mathematical transformation
    of data
transplantation 45
transverse studies 26, 27, 86
trauma *see* health of preparation;
    hormones
two-tailed statistical test 94, 98,
    (132)

type I error, type II error 21, 84,
    85, 93–100, (135, 195, 198)

units 58, 61–3, (132, 133, 143,
    144)
unmatched, unpaired data 91, 92,
    (136, 174–6, 195)
unpaired experimental designs *see*
    presentation of results;
    statistics
upgrading data 89–91

ventilation 33–5, (184, 185)

Wilcoxon matched pairs signed
    ranks test 83, 97, (174)